T0200743

BIOMIMETIC, BIORESPONSIVE, AND BIOACTIVE MATERIALS

BIOMIMETIC, BIORESPONSIVE, AND BIOACTIVE MATERIALS

An Introduction to Integrating Materials with Tissues

Edited by

Matteo Santin
Gary Phillips

Brighton Studies in Tissue-Mimicry and Aided Regeneration (*BrightSTAR*)
School of Pharmacy and Biomolecular Sciences
University of Brighton
Brighton, UK

A JOHN WILEY & SONS, INC., PUBLICATION

Published by John Wiley & Sons, Inc., Hoboken, New Jersey.
Published simultaneously in Canada.

For general information on our other products and services or for technical support, please contact our Customer Care Department within the United States at (800) 762-2974, outside the United States at (317) 572-3993 or fax (317) 572-4002.

Wiley also publishes its books in a variety of electronic formats. Some content that appears in print may not be available in electronic formats. For more information about Wiley products, visit our web site at www.wiley.com.

Library of Congress Cataloging-in-Publication Data:

Biomimetic, bioresponsive, and bioactive materials: an introduction to integrating materials with tissues/ edited by Matteo Santin, Gary Phillips.
    p. cm.
  Includes index.
  ISBN 978-0-470-05671-4 (hardback)
  1. Biomimetic polymers.   2. Biomimetics.   3. Tissues–Mechanical properties.   I. Santin, Matteo. II. Phillips, Gary.
    QD382.B47B56 2012
    660.6–dc23

                              2011021420

Printed in the United States of America.

10 9 8 7 6 5 4 3 2 1

CONTENTS

v

PREFACE

In the last 40 years, clinicians, industrialists, and patients have witnessed and experienced one of the most exciting advances achieved by modern science and technology; the research and development of medical devices. The coming together of bioengineering and materials science has played a fundamental role in the production of devices that are able to save the lives of many patients worldwide or significantly improve the quality of life for patients in those countries where the incidence of aging and lifestyle-related diseases have become a paramount social issue. Learning lessons from other fields of materials science, biomedical devices have been designed which are able to restore functionality in limbs and in the cardiovascular system as well as to replace the functions of compromised organs.

The relatively recent integration of molecular and cell biologists in the search for high-performance biomaterials and biomedical devices has led to the transformation of this research field from an engineering and materials science-based discipline into a truly interdisciplinary area of investigation. In the past few decades, the need to address the host response toward the implanted materials and to encourage tissue repair at their surfaces has sparked a new research approach in which biological processes have been studied in the presence of the challenge of an artificial surface. As a result, research worldwide has been driven by the search for biomaterials and devices specifically designed for targeted clinical applications, and new biochemical and cellular pathways have been identified.

An understanding of the finely tuned dependence of the activity of immunocompetent and tissue cells on the surrounding environment has led to a paradigm shift in the biomaterials field where the goal of tuning tissue response toward biomimetic, bioresponsive, and bioactive biomaterials has widely been accepted by the scientific community.

As a truly interdisciplinary community, we are now witnessing and experiencing a new era for our discipline where the concepts of biomimicry, bioresponsiveness, and bioactivity are associated not only to the production of new biomedical devices, but also to biomaterials able to drive the complete regeneration of tissues and organs, the integrity of which has been compromised by trauma, disease, or aging.

The present book aims to mark this era by illustrating the advances made thus far and critically discussing the challenges that still need to be faced. The book aims not only to provoke the thoughts of the experts, but also to stimulate a new generation of young students and scientists who will certainly be the protagonists of the future progress in this field. By presenting lessons from successful and unsuccessful stories and by critically assessing the state-of-the-art at research and clinical level, the editors of this book have aimed to provide the community with their contribution and to stimulate new research questions that will be able to open new routes of exploration.

The editors would like to express their gratitude to John Wiley & Sons for believing in this initiative and for supporting them throughout their editorial journey. The most profound gratitude goes to all those valuable colleagues who have given their expertise and availability to this project, thus making it possible; it is a further testimony to the value of many years of collaborations spent together in exciting research projects.

It is hoped that the reader of this book will find its reading a rewarding experience and appreciate its various sections as well as the colored illustration of the figures that can be accessed through ftp://ftp.wiley.com/public/sci_tech_med/biomimetic_bioresponsive.

MATTEO SANTIN
GARY PHILLIPS

CONTRIBUTORS

Mário A. Barbosa, Biomaterials Division, INEB-Instituto de Engenharia Biomédica, Universidade do Porto, Porto, Portugal

Roberto Chiesa, Dipartimento di Chimica, Materiali e Ingegneria Chimica, Politecnico di Milano, Milano, Italy

Alberto Cigada, Dipartimento di Chimica, Materiali e Ingegneria Chimica, Politecnico di Milano, Milano, Italy

Montserrat Espanol, Universitat Politècnica de Catalunya, Department of Materials Science and Metallurgical Engineering, Barcelona, Spain

Maria-Pau Ginebra, Universitat Politècnica de Catalunya, Department of Materials Science and Metallurgical Engineering, Barcelona, Spain

Paolo Tranquilli Leali, Department of Orthopaedic Surgery, University of Sassari, Sassari, Italy

Andrew L. Lewis, Biocompatibles UK Ltd, Farnham, Surrey, UK

Andrew W. Lloyd, Dean Faculty of Science and Engineering, University of Brighton, Moulsecooomb, Brighton, UK

Antonio Merolli, Orthopaedics and Hand Surgery, The Catholic University of Rome, Rome, Italy

Edgar B. Montufar, Universitat Politècnica de Catalunya, Department of Materials Science and Metallurgical Engineering, Barcelona, Spain

Román A. Pérez, Universitat Politècnica de Catalunya, Department of Materials Science and Metallurgical Engineering, Barcelona, Spain

Gary Phillips, Brighton Studies in Tissue-mimicry and Aided Regeneration (*Bright-STAR*), School of Pharmacy and Biomolecular Sciences, University of Brighton, Brighton, UK

Matteo Santin, Brighton Studies in Tissue-mimicry and Aided Regeneration (*Bright-STAR*), School of Pharmacy and Biomolecular Sciences, University of Brighton, Brighton, UK

Gabriela Voskerician, Department of Biomedical Engineering, Case Western Reserve University, Cleveland, OH, and Krikorjan, Inc., Menlo Park, CA, and Biodesign Innovation Group, Stanford University, Stanford, CA

1

HISTORY OF BIOMIMETIC, BIOACTIVE, AND BIORESPONSIVE BIOMATERIALS

Matteo Santin and Gary Phillips

1.1 THE FIRST GENERATION OF BIOMATERIALS: THE SEARCH FOR "THE BIOINERT"

Since it was first perceived that artificial and natural materials could be used to replace damaged parts of the human body, an "off-the-shelf" materials selection approach has been followed. These materials, now referred to as "first-generation" biomaterials, tended to be "borrowed" from other disciplines rather than being specifically designed for biomedical applications, and were selected on the basis of three main criteria: (1) their ability to mimic the mechanical performances of the tissue to be replaced, (2) their lack of toxicity, and (3) their inertness toward the body's host response [Hench & Polack 2002].

Following this approach, pioneers developed a relatively large range of implants and devices, using a number of synthetic and natural materials including polymers, metals, and ceramics, based on occasional earlier observations and innovative approaches by clinicians. Indeed, many of these devices are still in use today (Figure 1.1A–J). A typical example of this often serendipitous development process was the use of poly(methyl methacrylate) (PMMA) to manufacture intraocular and contact lenses. This material (Table 1.1) was selected following observations made by the clinician Sir Harold Ridley that fragments of the PMMA cockpit that had penetrated into

Biomimetic, Bioresponsive, and Bioactive Materials: An Introduction to Integrating Materials with Tissues, First Edition. Edited by Matteo Santin and Gary Phillips.
© 2012 John Wiley & Sons, Inc. Published 2012 by John Wiley & Sons, Inc.

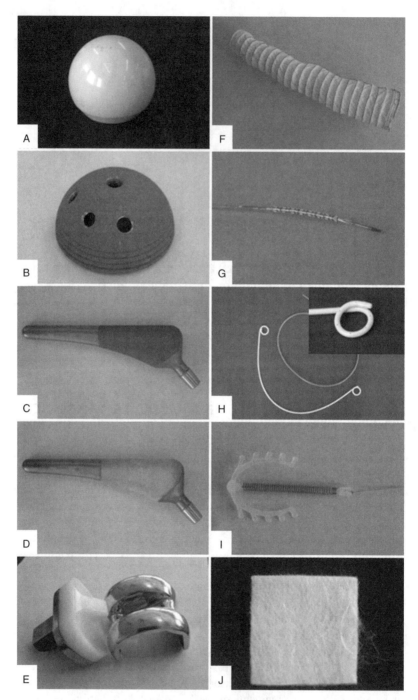

Figure 1.1. Examples of medical implants and their components. (A–F) Orthopedic implant components: (A) femor head, (B) hip socket, (C) titanium stent coated with porous titanium foam, (D) titanium stent coated with hydroxapatite coating, (E) knee implant components. (F–J) Other types of biomedical implants: (F) vascular graft, (G) coronary stent, (H) ureteral stent (insert shows detail of the device pig-tail end), (I) intrauterine device, (J) wound dressing.

TABLE 1.1. Chemical Structure of Typical Polymeric Biomaterials

Polymer	Structure
Polystyrene	
Poly(vinyl chloride)	
Poly(2-hydroxyethyl methacrylate)	
Poly(methyl methacrylate)	
Chronoflex 80A poly(urethane) Hydrothane poly(urethane) PFPM/PEHA75/25 Poly(octofluoropentyl methacrylate/ethylhexyl methacrylate)	-(-PHECD-DesmW-BD-)-$_n$ Polyethylene glycol polyurethane

the eyes of World War II pilots induced a very low immune response (see Section 1.3.1) [Williams 2001].

Also against the backdrop of the Second World War, a young Dutch physician named Willem Kolff pioneered the development of renal replacement therapies by taking advantage of a cellophane membrane used as sausage skin to allow the dialysis of blood from his uremic patients against a bath of cleansing fluid [Kolff 1993]. Later, in the early 1960s, John Charnley, learning about the progress materials science had made in obtaining mechanically resistant metals and plastics, designed the first hip joint prosthesis able to perform satisfactorily in the human body [Charnley 1961]. These are typical examples of how early implant materials were selected; however, it was soon

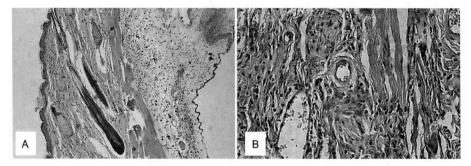

Figure 1.2. Fibrotic capsule formed around a commercial polyurethane biomaterial: (A) general view, (B) high magnification showing histological details of the fibrotic tissue.

recognized that the performance of these materials was often limited by the host response toward the implant, which often resulted in inflammation, the formation of a fibrotic capsules around the implant, and poor integration with the surrounding tissue (Figure 1.2A,B) [Anderson 2001].

The poor acceptance and performance of early biomaterials indicated that their interaction with body tissues was a complex problem that required the development of more sophisticated products. As a result, it was realized that inertness was not a guarantee of biocompatibility. Indeed, in 1986, the Consensus Conference of the European Society for Biomaterials put in place widely accepted definitions of both biomaterials and biocompatibility, which took into account the interaction between an implanted material and biological systems. According to these definitions, a biomaterial was "a nonviable material used in a medical device intended to interact with biological systems," whereas biocompatibility was defined as "the ability of a material to perform with an appropriate host response in a specific application" [Williams 1987]. Perhaps for the first time, materials scientists and clinicians had an agreement on what their materials should achieve. However, as you will see later in this chapter and throughout this book, the ever-expanding fields of biomaterials science and tissue engineering call for newer and more specific definitions.

1.1.1 Bioinert: Myth, Reality, or Utopia?

In the 1980s, the formation of a fibrotic capsule "walling off" many biomedical implants from the surrounding tissue triggered biophysical and immunological studies that identified the molecular, biochemical, and cellular bases of the host response that caused the formation of this interposed and pathological tissue [Williams 1987; Anderson 1988]. In particular, many studies highlighted that this host response could not be avoided due to the immediate deposition of proteins onto the material surfaces and their change of conformation [Norde 1986]. The material surface-induced conformational changes transformed the host proteins into foreign molecules, antigens, which were capable of eliciting a foreign body response by the host (Figure 1.3).

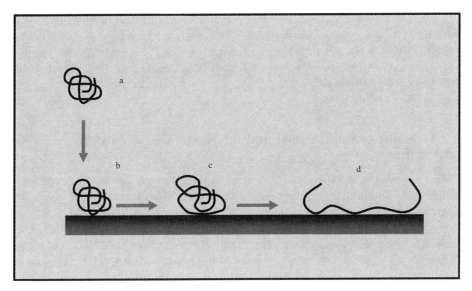

<u>Figure 1.3.</u> Schematic representation of protein denaturation upon adsorption on biomaterial surface: (a) soluble protein approaches biomaterial surface, (b) protein adsorbs to material surface, (c) protein starts to unfold through interactions with material surface, (d) protein acquires an antigenic conformation.

Biomaterial surfaces contacted by blood, saliva, urine, cerebrospinal and peritoneal fluids, or tears cannot avoid interactions with proteins and other molecules that are naturally contained in the overlying body fluid [Santin et al. 1997]. As a consequence, the implant surface is rapidly covered by a biofilm that masks the material surface and dictates the host response. It is clear, therefore, that as a result of these processes, no biomaterial can be considered to be totally inert. However, although they are subjected to a continuous turnover, it is a fact that proteins (and more broadly, all tissue macromolecules) retain their native conformation during the different phases of tissue formation and remodeling. Hence, for the past two decades the scientific community has striven for the development and synthesis of a new generation of biomaterials that are able to control protein adsorption processes and/or tissue regeneration around the implant.

1.2 THE SECOND GENERATION OF BIOMATERIALS: BIOMIMETIC, BIORESPONSIVE, BIOACTIVE

In conjunction with the findings regarding the biochemical and cellular bases of host response toward implants, material scientists began their search for biomimetic, bioresponsive, and bioactive materials capable of controlling interactions with the surrounding biological environment and that could participate in tissue regeneration processes.

The move toward the synthesis and engineering of this type of biomaterial was initiated by the discovery of ceramic biomaterials that were proven to favor the integration of bony tissue in dental and orthopedic applications [Clarke et al. 1990], as well as by the use of synthetic or natural polymers [Raghunath et al. 1980; Giusti et al. 1995]. Second-generation metals also emerged that were able to improve the integration with the surrounding tissue.

1.2.1 Hydroxyapatite (HA) and Bioglass®: Cell Adhesion and Stimulation

The ability of HA, and Bioglass to integrate with the surrounding bone in orthopedic and dental applications strictly depends on the physicochemical properties of these two types of ceramics, which will be described in Chapter 7. Here, it has to be mentioned that since their early discovery and use in surgery, the good integration of these biomaterials with bone, the osteointegration, depends on mechanisms of different nature that have triggered new concepts/definitions and new technological targets among scientists.

Although HA osteointegration can intuitively be attributed to their ability to mimic the bone mineral phase (see Chapter 3), the mechanisms underlying Bioglass-induced bone formation are not as clearly identifiable. It is known that HA favors the deposition of new bone on its surface by supporting osteoblast adhesion and by promoting the chemical bonding with the bone mineral phase [Takeshita et al. 1997]. Furthermore, the ability of HA to induce bone formation when implanted intramuscularly in animals, allegedly via the differentiation of progenitor cells, clearly shows their osteoinductive potential; indeed, osteoinductivity is defined as the ability of a biomaterial to form ectopic bone. Conversely, Bioglass osteoinductivity seems to be intrinsic to its degradation process whereby (1) growth factors remain trapped within the gel phase formed during the degradation of the material and, consequently, released to the cells upon complete material dissolution; (2) structural proteins of the extracellular matrix (ECM) such as fibronectin form strong bonds with particles of the degrading material; and (3) silicon ions stimulate osteoblast (and allegedly progenitor cell) differentiation and, subsequently, the production of new bone [Xynos et al. 2000]. Regardless of the type of ceramic used, it is now widely recognized that the topographical features of these types of biomaterials are also fundamental to their bioactivity. For example, the absence of porosity or porosity of different sizes may lead to no osteointegration or to only poor bone formation [Hing et al. 2004].

1.2.2 Collagen, Fibrin Glue, and Hyaluronic Acid Hydrogels: Presenting the ECM

The use of collagen, fibrin, and hyaluronan, which are all natural components of the ECM, was born from scientists' intuition that tissue cells recognize these biopolymers as natural substrates to form new tissue.

Fundamental to the application of these biological materials was an appreciation of their physicochemical and biological properties. Collagen is the most ubiquitous structural protein in the human body and the principal constituent of ECM in connective tissues [Rivier & Sadoc 2006]. It consists of a tightly packed structure composed of three polypeptide chains that wind together to form a triple helix [Rivier & Sadoc 2006]. These collagen molecules then associate to form collagen fibrils. A number of reviews are available on the structure of the different types of collagen found throughout the body [Engel & Bachinger 2005; Rivier & Sadoc 2006]. Collagen plays a key role in the wound healing process and the development of cartilage and tendons, and it is known that collagen can favor the formation of HA on its structure, thus inducing bone mineralization [Zhai & Cui 2006]. As part of the ECM, collagen provides a suitable milieu for cell proliferation, migration, and differentiation during the production of new tissue via its biodegradation and tissue remodeling. Collagen is, therefore, a natural biomaterial whose inherent potential has been exploited by biomaterials scientists in ligament replacement and other tissue engineering applications [Rothenburger 2001; Gentleman et al. 2003, 2006; Boccafoschi 2005; Kutschka et al. 2006], and collagen types I and IV have been commercialized as dermal substitutes [Jones et al. 2006].

The use of fibrin as a biomaterial was founded on the fact that fibrin clots are self-assembling networks with biological and physicochemical attributes that have the potential to be used in a number of biomedical applications. Three-dimensional (3D) porous fibrin networks are formed through a series of events during the blood coagulation cascade, resulting in the formation of a biopolymer gel material. The structure of the gel is determined by the thrombin-mediated conversion of fibrinogen to fibrin and the subsequent self-assembly of the fiber network [Helgerson et al. 2004]. Fibrin glue saw its first application as a surgical adhesive, but in the emerging era of tissue engineering, it has been suggested by many scientists as a suitable gel for cell encapsulation (Figure 1.4a–c) [Bach et al. 2001]. This is due to the fact that fibrin clots provide a structural scaffold that allows the adhesion, proliferation, and migration of cells important in the wound healing process and, when associated with proteins as a clot, has intrinsic biological properties that support and control, to some extent, cell differentiation. Fibrin-based biomaterials also benefit from the fact that they are naturally remodeled and resorbed as part of the fibrinolytic processes associated with the cellar deposition of a new ECM as part of the normal wound healing processes [Helgerson et al. 2004] (see Section 1.3.1).

Hyaluronan, one of the main components of cartilage (see Chapter 3), has been chemically modified and commercialized to favor cartilage and skin regeneration (see Chapter 6) [Barbucci et al. 1993]. Hyaluronan consists of a single polysaccharide chain with no peptide in its primary structure, and it has a molecular weight that reaches millions of Daltons [Fraser et al. 1997]. The biological properties of this molecule are imparted by specific hyaluronan binding sites present in other ECM molecules and on the surface of cells [Fraser et al. 1997]. A number of proteins exist—the hyaladherins—that have the ability to recognize hyaluronan and result in the binding of hyaluronan molecules with proteoglycans to reinforce the structure of the ECM [Fraser et al. 1997; Day & Prestwich 2002]. At the molecular and cellular levels, it is now known that these biomolecules are able to support tissue regeneration because of the presence of specific bioligands that are able to recognize receptors on the cell membrane which, in turn, stimulates cell functions [Turley et al. 2002] (see Section 1.3.1.1).

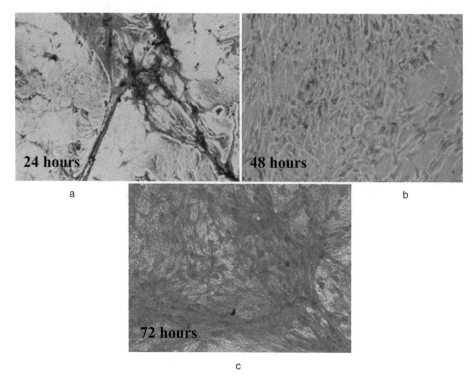

Figure 1.4. Collagen deposition by osteoblasts encapsulated in a fibrin hydrogel. Incubation times: (a) 24 hours, (b) 48 hours, and (c) 72 hours.

Furthermore, the physicochemical and biochemical properties of the three molecules discussed here can favor the interaction with other tissue components, forming organized macromolecular structures capable of conferring on tissues their specific mechanical properties (see Chapter 2). As previously mentioned, it is known that collagen can favor the formation of HA on its structure, thus inducing bone mineralization [Zhai & Cui 2006], and that hyaluronan is capable of interacting with other proteins to form macromolecular structures that are able to retain a relatively high water content. This high water content thus acts as an effective shock absorber in cartilage and ocular tissues [Fraser et al. 1997]. However, these substrates have shown some drawbacks and limitations. Although they provided the regenerating tissue with some important properties, some others were missing. As mentioned above, one of the main benefits of using these biopolymers in clinical applications is that they promote biorecognition. However, in most cases, this biorecognition is not specific for the type of cells that need to be targeted to induce tissue regeneration. For example, collagen and fibrin, as well as other important ECM proteins (e.g., fibronectin), present in their structure the arginine–glycine–aspartic acid (RGD) sequence that is recognized by most tissue cells as well as by inflammatory cells such as monocytes or macrophages [Phillips & Kao 2005]. As a result of this relatively broad spectrum of cell recognition, collagen-based biomaterials have been shown to induce an immune response in patients, which often leads to the formation of fibrotic tissue (see Section 1.1). In addition, collagen-based implants, either extracted from mammalian sources or from recombinant bacteria, may not represent the composition of the real ECM and miss some components required to regulate the process of tissue regeneration. For example, it has been proven that physiological skin ECM collagen presents, on its surface, proteins such as α1-microglobulin, which is capable of modulating the activity of resident macrophages [Santin & Cannas 1999]. The absence of this protein in pathological tissues (e.g., scar tissue) and collagen implants seems to lead to a collagen-induced activation of immunocompetent cells. The modulating action of the α1-microglobulin is likely to be only one aspect of a multi-faceted process leading to the regulation of the immunocompetent cell activity in connective tissues. Therefore, collagen-based implants, although representing a step forward in developing biomaterials for tissue regeneration, address the problem in a relatively simplistic manner. A plethora of immunomodulators are present in physiological tissues, which may need to be taken into account to improve the performance of the collagen-based biomaterials.

Similarly, hyaluronan is recognized by cell receptors such as CD44, which are present on the membrane of both tissue and inflammatory cells. The role of this polysaccharide in nature is tuned by its molecular weight [Mytar et al. 2001; Teder et al. 2002]. It has been proven that low molecular weight hyaluronan is fundamentally proinflammatory and angiogenic, thus promoting the formation of granular tissue. Conversely, relatively high molecular weight hyaluronan seems to prevent angiogenesis and inflammation. Thus far, at the clinical level, relatively high molecular weight hyaluronan and its ester derivatives have been used, but not enough information has been collected to optimize the molecular weight of this polysaccharide. More accurate studies may be able to define the appropriate physicochemical characteristics of hyaluronan-based biomaterials to encourage some degree of vascularization and inflammation, which are required for a physiological regeneration.

Finally, although the use of fibrin glue as an adhesive material in surgery is widespread and successful, the tissue regeneration potential of this natural hydrogel has been proven to be limited unless key growth factors are loaded in its mesh. As for collagen and hyaluronan, this is not surprising considering that the main function of fibrin is to stop the bleeding and provide the damaged tissue with a temporary scaffold for its repair.

Each of the biopolymers mentioned in this section have reached the market and provided good, although not always satisfactory, clinical performances. Nevertheless, the use of these materials in clinics has opened the door to the development of biomimetic biomaterials able to mimic the structure, biochemistry, and biofunctionality of tissue components.

1.2.3 Chitosan and Alginate: Replacing the ECM

As previously mentioned, the ECM is a structural, 3D network consisting of a number of macromolecules and polyelectrolytes including fibronectin, proteoglycan, collagen, laminin, and glycosaminoglycans. This macromolecular architecture mediates the interaction of cells with the substrate and provides a scaffold for cell migration and proliferation [Zaidel-Bar et al. 2004]. In addition to using molecules that naturally occur as components of the ECM, a number of attempts have been made to replace this scaffold using polymers from other natural sources either individually or in polyelectrolyte complexes to form hydrogels or solid porous constructs [Hayashi 1994; Madihally & Matthew 1999]. Two of the principal macromolecules used for these applications are chitosan, the deacetylated product of chitin from the exoskeleton of shellfish and alginate, derived from brown algae, both of which have been used as biodegradable materials for wound healing, tissue reconstruction, cell encapsulation, and drug delivery [Tomihata & Ikada 1997; Madihally & Matthew 1999; Jayakumar et al. 2006; Roughley et al. 2006]. The use of these polymers, either individually or in combination with others to support and reinforce the regenerating tissue, have underpinned the ever-expanding discipline of tissue engineering [Minuth et al. 1998; Madihally & Matthew 1999].

However, although generally accepted to have favorable biocompatibility and toxicity profiles (Rao & Sharma 1997), it has been reported that chitosan polymers used as soluble polymeric carriers for intravenous administration or following particulate degradation may induce cellular toxicity (Carreño-Gómez & Duncan 1997). More recent studies have suggested that hydrogel scaffolds containing collagen, chitosan, and HA elicit a severe inflammatory response associated with an inadequate ingrowth of neovascularization from the surrounding host tissue when implanted in dorsal skinfold chambers of mice [Rücker et al. 2006]. It is apparent, therefore, that as the applications for these materials are explored, their biocompatibility may be altered, depending on the situation.

Like chitosan, alginate is a natural polymer that can be prepared on its own into a number of physical forms, including beads for cell encapsulation and porous sponges suitable for cell ingrowth and neovascularizaton. The materials produced are relatively nontoxic and noninflammatory, although their applications tend to be limited due to poor mechanical properties and cell performance. These shortcomings have been addressed by combining the alginate with other materials including chitosan (Rosca et al. 2005).

1.2.4 Poly(Lactic/Glycolic) Acid Copolymers: Encouraging Tissue Remodeling by Safe Biodegradation

The development of a second generation of biomaterials found its inspiration in nature, not only by trying to mimic the biochemical and structural features of natural tissue, but also by taking into account its ability to undergo resorption during the physiological turnover typical of the tissue remodeling processes (see Chapters 2 and 3). In the 1980s, materials scientists recognized the importance of biodegradation in allowing tissue ingrowth. Thus far, biodegradable biomaterials, although ideal for tissue regeneration, have been confined to the manufacture of implants not requiring load-bearing capabilities. In an attempt to synthesize polymers that are able to biodegrade at a rate tuned with tissue regeneration and to ensure the release of degradation by-products that are not toxic for the host, biomaterials based on natural molecules such as lactic and glycolic acid have been developed. Materials scientists have exploited methods of syn-

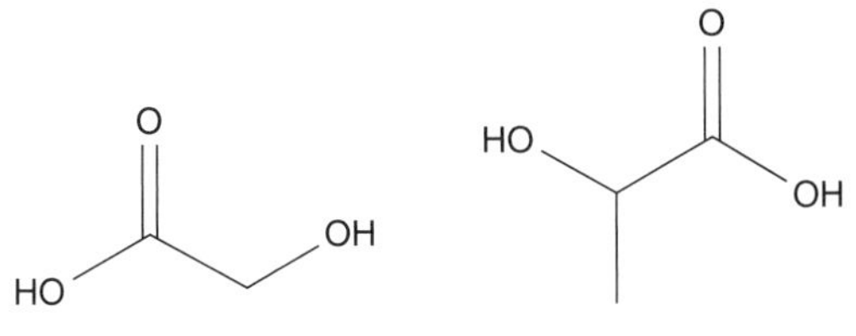

Figure 1.5. L-lactic and L-glycolic acid monomers utilized for the synthesis of poly(lactic/glycolic) acid copolymers.

thetic chemistry to produce polymers of these natural molecules and combinations of the two in the form of copolymers (Figure 1.5) [Grayson et al. 2004]. It has since been demonstrated that these polymers can degrade into very basic molecular species such as CO_2 and H_2O, thus ruling out the formation of any toxic by-product. In addition, the combination of the two monomers in different proportions in copolymer formulations allows the tuning of their degradation rate, depending on the required biomedical application. However, relatively recent studies have demonstrated that before complete dissolution, fragments of these polymers elicit an inflammatory response, thus altering tissue regeneration [Grayson et al. 2004].

1.2.5 Porous Metals: Favoring Mechanical Integration

The transition that took place in the 1980s with the movement toward second-generation biomaterials also involved metals. During the past four decades and, indeed, to the present day, materials scientists and biomedical companies have been facing the need to provide clinicians with implants that are able to sustain relatively high and protracted biomechanical stresses. In most cases, these stresses cannot be sustained unless metals are used in the implant manufacture. However, ensuring the integration of metal implants into tissues remains a significant challenge. A major step forward toward this objective has been the improvement of both device design and surface properties, the former leading to biomechanically performing implants, the latter to mechanical integration with the surrounding tissue [Takemoto et al. 2005]. Indeed, the improved distribution of mechanical loads transferred to orthopedic and dental implants has reduced mechanical stresses on the tissue/implant interface and the consequent failure of the implants caused by stress-induced bone fractures. The introduction of surface porosity has led to an enhanced grip of the tissue during its growth around metal implants. At the cellular level, it has been proven that a rough surface can improve cell adhesion to metal implants and, as a consequence, their colonization of the implant surface (Figure 1.6) [Sandrini et al. 2005].

As a result of the development of improved metal implants, the thick fibrotic capsule that typically formed around the first generation of metal orthopedic and dental

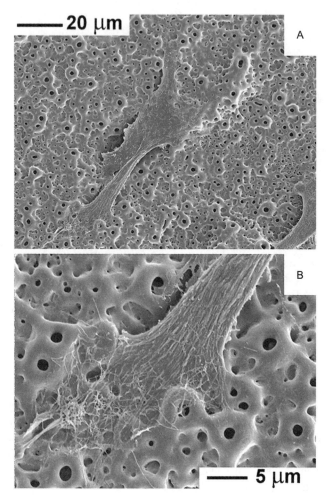

Figure 1.6. Osteoblast adhesion on rough implant surface: (A) adhering osteoblast, (B) osteoblast focal adhesion establishing contact with rough surface and secretion of collagen fibrils.

implants has been reduced to a thin layer of soft tissue interposed between the metal surface and the mineralized tissue (Figure 1.7) [Steflik et al. 1998]. The integration of these implants has thus been significantly improved and their clinical life extended, but mechanical failure is still the destiny of most of these medical devices.

Chapters 3 and 6 of this book will demonstrate how the introduction of biomimetic, bioactive, and bioresponsive functionalization methods promises to lead to metal implants with improved biological performances under biomechanical loads.

Titanium implant surface

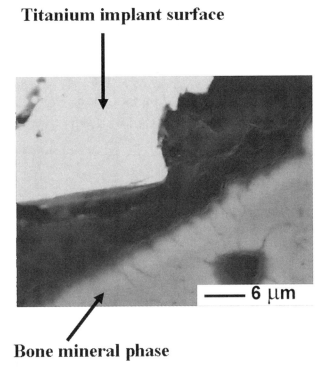

— 6 μm

Bone mineral phase

Figure 1.7. Bone regeneration around a porous titanium implant. Back-scattered scanning electron micrograph showing the implant surface separated from bone by a nonmineralized area.

1.3 THE THIRD-GENERATION BIOMATERIALS: BIOMIMICKING NATURAL BIOACTIVE AND BIORESPONSIVE PROCESSES

In the 1990s, it was evident that a third-generation of biomaterials was required that was capable of improving the clinical performance of implants by harnessing their potential to interact with surrounding tissues. As a consequence, new technological advances were advocated that would fulfill the ambition of abandoning the clinical approach of tissue replacement and achieving tissue regeneration. Tissue regeneration is, therefore, a requirement for both the integration of permanent implants and for a complete tissue regeneration supported by biodegradable biomaterials. It was envisaged that the bioactivity of ceramics and natural polymers could be mimicked by the synthesis of new biomaterials, simultaneously offering adequate physicochemical properties, biointegration potential, and ease of handling during surgical procedures.

Third-generation biomaterials have been designed to modulate processes that are fundamental to tissue regeneration, including cell adhesion, proliferation, and differentiation through the activation of particular genes [Hench & Polack 2002]. Biomimetic and bioactive biomaterials have been synthesized, which are able to target specific mechanisms of cell adhesion with the ultimate goal of regenerating tissues such as bone, cartilage, vascular tissue, and nerves. Their mechanisms of action can only be understood by learning the basics of the biological processes that they mimic.

1.3.1   Principal Phases of Tissue Regeneration

Tissue regeneration takes place through four main phases (Figure 1.8a–d) [Martin & Leibovich 2005]:

1. Clot formation
2. Inflammatory response
3. Cell migration/proliferation
4. New ECM deposition

The formation of a clot is required to stop bleeding following trauma, whether accidental or as a consequence of a surgical procedure. The clot is formed through the activation of a blood plasma protein (fibrinogen), which is transformed into a polymeric form (fibrin) by an enzyme called thrombin and cross-linked by a transglutaminase enzyme called Factor XIII (Figures 1.8a and 1.9a). Platelets, which are responsible for fibrinogen activation, complete the plug by being entrapped in the fibrin mesh (Figure 1.9b). Platelets soon degranulate, releasing a series of biochemical signals that recruit tissue and inflammatory cells to the site of injury. Gradually, the fibrin mesh is invaded by inflammatory cells such as the neutrophils (or polymorphonucleate granulocytes) and, later, by monocytes/macrophages, which clear tissue debris and infiltrated bacteria from the site, and begin to digest the fibrin clot (Figure 1.8b). Monocytes/macrophages

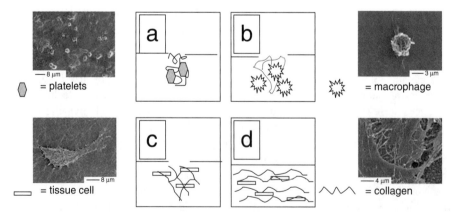

Figure 1.8. Principal phases of tissue regeneration: (a) clot formation, (b) clot infiltration by inflammatory cells, (c) tissue cell migration, (d) ECM deposition. Inserts show micrographs of typical cell morphologies.

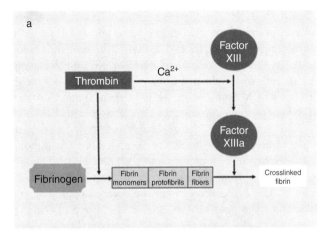

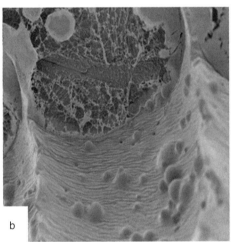

Figure 1.9. Fibrin clot formation: (a) schematic representation of the biochemical pathway, (b) scanning electron micrograph of a clot formed on the surface of a vascular graft.

also play a central role in tissue regeneration by releasing important biochemical signals called cytokines and growth factors, which modulate the inflammatory response (see Section 1.1.) as well as the formation of new tissue. This new tissue formation takes place via an early phase of cell migration and proliferation (Figure 1.8c) followed by the synthesis and deposition of a new ECM composed mainly of structural proteins (e.g., collagen, fibronectin, and laminin) and glycosaminoglycans (Figure 1.8d) [Turley 2001]. All of these phases take place with some degree of overlap and are ultimately dependent on the activation of cells and their constant interaction with the surrounding matrix and other cells. Indeed, cell activation is modulated not only by the biochemical signaling constituted by cytokines and growth factors but also through the adhesion to components of the ECM.

1.3.1.1 Cell Adhesion: The Cornerstone of Tissue Regeneration. During the very early phases of their activation, tissue cells synthesize and secrete hyaluronan into the pericellular space [Zaidel-Bar et al. 2004]. This proteoglycan-based halo fills the gap that exists between the cell membrane surface and the components of the damaged ECM. Because of its physicochemical properties, hyaluronan can establish interactions with other ECM components while being recognized by the cell through a specific class of membrane receptors called CD44 [Fraser et al. 1997]. This type of cell adhesion, also called "soft contact" [Zaidel-Bar et al. 2004], occurs as soon as the cell faces the new environment. It is replaced after a few seconds by a more stable contact mediated by a network of anchor proteins on the cell membrane. These proteins cluster together in various arrays to form anchoring patterns: focal adhesion, fibrillar adhesion, and focal complexes. These anchoring patterns are very important as they dictate cell motility and, therefore, its migration to the wound site. The anchoring proteins responsible for the adhesion of cells to the pericellular matrix and to neighboring cells are a class of membrane receptors called integrins. These are heterodimeric proteins composed of an α subunit associated through noncovalent interactions to a β subunit [Stefansson et al. 2004] (Figure 1.10). The two types of subunits are separate families of proteins: The α subunit includes 18 different members, while the β subunit accounts for 8 different proteins. Both the subunits span across the cell membrane and can, therefore, be schematically divided into three main domains (Figure 1.10):

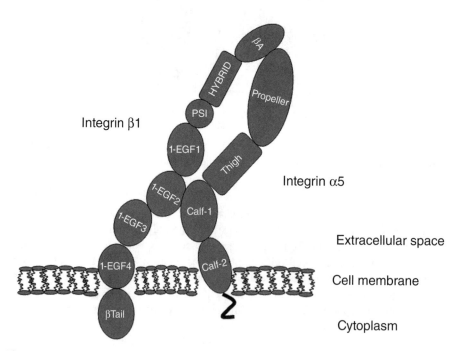

<u>Figure 1.10.</u> Schematic representation of a typical integrin showing the α and β subunits and their localization across the cell membrane.

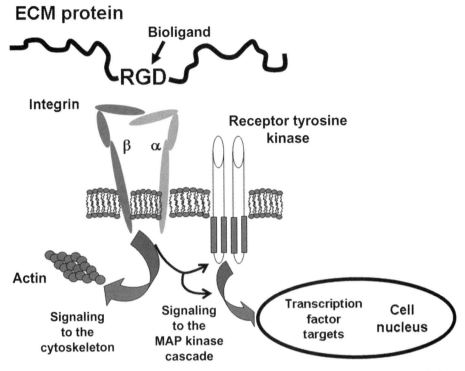

Figure 1.11. Schematic representation of the relationship between ECM bioligands (i.e., RGD), integrin, and intracellular signaling.

(1) N-terminal extracellular domain
(2) Transmembrane domain
(3) Cytoplasmic domain

Integrins are important, not only for cell anchoring to the surrounding extracellular environment through their interaction with specific bioligands, but also because of their ability to generate intracellular signaling (Figure 1.11). This signaling modulates the main cell activities related to tissue regeneration such as cell migration, proliferation, and differentiation. These activities have been defined as "outside-in signaling," which is the ligand-induced signaling mechanism and "inside-out signaling," which includes a series of mechanisms that activate the integrins.

Integrin activation is determined by conformational changes and clustering of the subunits and of their respective domains. Each one of these modifications seems to lead to particular cell activation pathways. In particular, the activation of the transmembrane domain is believed to contribute to signaling in both directions: outside-in and inside-out. The interactions of bioligands exposed on the ECM structure with integrins leads to the activation of the so-called focal adhesion kinase (FAK) and of other not fully characterized tyrosine kinases. Once a cell has adhered to ECM components via bioligand/integrin interactions, the autophosphorylation of these kinases takes place by integrin clustering and interactions between the cytoplasmic domain of the β subunits and the actin filaments [Juliano et al. 2004; Li et al. 2004]. The interactions with the actin filaments take place through the recruitment of cytoplasmic proteins and by a plethora of downstream responses, which leads to the remodeling of the actin filaments and ultimately to cell motility [Juliano et al. 2004]. Because of these interactions, FAK, their downstream targets, and phosphatidylinositol 3-kinase have been shown to be important mediators of cell migration [Armulik et al. 2004]. Indeed, cell migration can be considered as one of the earliest events in tissue regeneration by which cells try to colonize the remodeling or damaged tissue to synthesize and deposit new ECM (Figure 1.8c). However, cell migration is also favored by the secretion of particular classes of enzymes, the metalloproteinases, which cleave specific peptide sequences in the ECM mesh, thus opening the space to the migrating cell [Buhling et al. 2006].

FAK phosphorylation in turn leads to the activation of another enzyme, the mitosis activation phosphorylase kinase (MAPK). Mitosis is the process by which cells divide and proliferate. Therefore, the activation of MAPK through integrins leads to increased cell proliferation, a key event in tissue regeneration. The effect of integrins on cell proliferation also includes their ability to undergo partner assembling with growth factor receptors. Increasing evidence suggests that the combined actions of integrins and receptor protein tyrosine kinases transduce proliferative signals in cells through an enzymatic cascade (Figure 1.11) [Cabodi et al. 2004; Juliano et al. 2004].

The effects of integrins are not limited to cell migration but also involve their proliferation and differentiation processes as well as the organization of ECM components [Velling et al. 2002; Boulter & van Obberghen-Schilling 2006]. In fact, it has been demonstrated that the polymerization and assembly of ECM proteins such as fibronectin, laminin, and collagens type I and III are enhanced by their interactions with integrins such as $\alpha_{11}\beta_1$ and $\alpha_2\beta_1$ and by cytoskeleton-associated tensions.

The interactions mediated by the integrins rely mainly on the presence of a specific amino acid sequence, the RGD sequence, in the structure of many ECM proteins [Takagi 2004]. This ligand is present in fibrinogen, fibronectin, vitronectin, and other ECM glycoproteins with a relatively high molecular weight.

Recently, the increased expression of integrin $\alpha_v\beta_5$ has been shown to induce the differentiation of dermal fibroblasts, which is driven by the colocalization of specific growth factor receptors such as the TGF-β1 receptor on the cell membrane [Asano et al. 2006]. Conversely, the fibronectin receptor integrin $\beta_1\alpha_5$ has been shown to be downregulated during the differentiation of mesenchymal stem cells into chondrocytes [Goessler et al. 2006]. These data seem to suggest that the role of integrins in cell differentiation is very complex and depends on the types of integrin, cell, and differentiation stage analyzed.

However, ECM modulates cell behavior by a plethora of bioligands that extend beyond hyaluronan and the RGD domain. Cell–cell and cell–ECM interactions also take place through lectins, a class of proteins able to bind carbohydrates both specifically and noncovalently [Sharon & Lis 1993]. Among them, the family of the galectins has been shown to be involved in a number of cellular processes. These span from cell adhesion [Mahanthappa et al. 1994] and the regulation of the cell cycle [Wells & Malluchi 1991] to cell differentiation [Goldrin et al. 2001], and endothelial cell motility and angiogenesis [Fukushi et al. 2004]. Other important peptide sequences have been found to modulate specific cell types and their relative functions. Table 1.2 lists those sequences more relevant to biomimetic biomaterials (see Section 1.4.1).

1.3.1.2 Mechanisms of Tissue Mineralization. Bone and cartilage have developed their ability to sustain biomechanical stresses by evolving as mineralized structures. The mineralization of these tissues is also driven by cell activation and is, therefore, strictly dependent on the mechanisms described in Section 1.3.1.1. For example, adhering and differentiating osteoblasts express enzymes (such as alkaline phosphatase), which facilitate the formation of a mineral phase. In addition, they secrete collagen and other calcium-binding proteins, and send signals to sister osteoblasts (autocrine signaling) and to other cells such as the osteoclasts (paracrine signaling) (Figure 1.12). The modulation of osteoblasts is key factor in the process of tissue resorption during bone remodeling [Buckwalter et al. 1995].

TABLE 1.2. Main Cell Binding Domains Utilized to Functionalize Biomimetic Biomaterials (Rezania & Healy 1999)

RGD	Cell adhesion domain of most ECM proteins
FHRRIKA	Osteoblast migration domain
YIGSR	Laminin cell binding domain
IKVAV	Laminin cell binding domain
REDRV	Fibronectin cell binding domain
LDV	Fibronectin cell binding domain
DGEA	Collagen type I cell binding domain

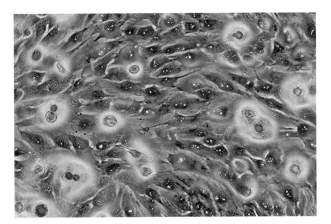

Figure 1.12. Typical plurinucleated osteoclasts differentiated from blood mononuclear cells under osteoblast-secreted stimuli. Osteoclast cells are identified as large, multinucleated cells within a prevalent osteoblast population.

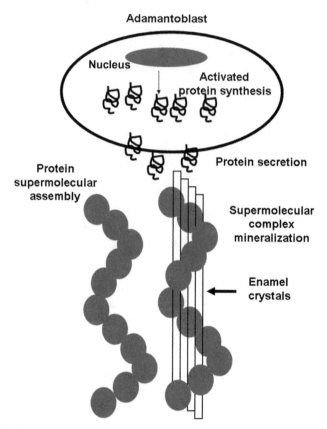

Figure 1.13. Schematic representation of amelogenin synthesis, secretion, supramolecular assembling, and mineralization. Adamantoblasts are cells producing enamel in teeth.

At the molecular level, the formation of an apatite-rich mineral phase depends on specific functional groups exposed on the surface of ECM components. Collagen and other calcium-binding proteins associated with the collagenic template are able to catalyze tissue mineralization [Camacho et al. 1999]. Amelogenins—globular proteins of tooth enamel—are capable of self-assembling into spiral-like structures, exposing their core calcium-binding functional groups and directing crystal growth to form one of the strongest mineral tissues in nature (Figure 1.13) [Fincham & Simmer 1997].

Another natural approach toward tissue mineralization is the phospholipid-mediated nucleation of apatite crystals [Boskey 1978]. Matrix vesicles (MVs) have been found in developmental bone and cartilage as well as in regenerating bone. The nature of MVs is controversial as it is not clear whether they are membrane fragments derived from apoptotic cell ghosts or if they are produced by osteoblasts through a specific mechanism. Regardless of their origin, it is known that MVs can facilitate the influx of calcium into their interior through calcium channels usually present in the cell membrane. Once inside the MV, calcium can complex with phosphorus ions through the catalytic

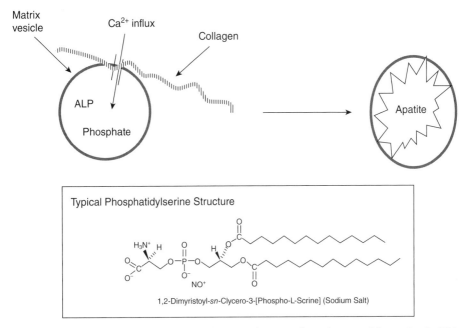

Figure 1.14. Schematic representation of the mechanism of apatite crystal formation in MV. Insert highlights the PS zwitterionic group catalyzing the crystal seeding process.

action of the zwitterion moiety of phosphatidylserine (PS), a membrane phospholipid (Figure 1.14).

The physicochemical mechanism by which these molecules induce mineralization of tissues is a strategy common to all mineralized tissue throughout the animal kingdom [Mann 1988] (Figure 1.15). This mechanism relies on the ability of certain biomolecules to expose functional groups with a high binding capacity but lower binding affinity for calcium ions. Because of their physicochemical properties, these groups allow the formation of a calcium coordination sphere (the anionic sphere) on the molecule surface and make it available to interact with phosphate ions (the cationic sphere). The presence of moieties with high calcium binding capacity and low binding affinity is indeed essential for the organization of the counterion clusters and subsequent apatite crystal formation. Conversely, a high calcium binding affinity would subtract calcium from phosphorus, making the reaction unfavorable. Once ion clusters are formed, crystals nucleate through the formation of amorphous calcium phosphate and eventually loose water molecules to become mature and water-insoluble apatite crystals.

1.4   PRINCIPLES OF BIOMIMESIS AND BIOACTIVITY

The third-generation biomaterials set the stage for a new era in biomedicine; a paradigm shift was seen where the concept of tissue replacement has gradually been replaced by

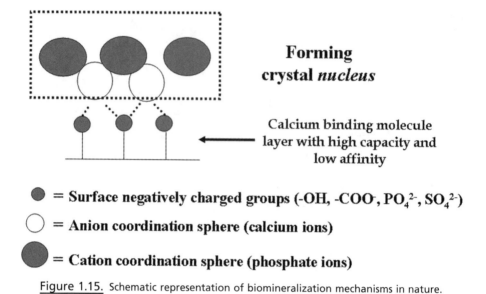

Forming crystal *nucleus*

Calcium binding molecule layer with high capacity and low affinity

● = Surface negatively charged groups (-OH, -COO⁻, PO₄²⁻, SO₄²⁻)

○ = Anion coordination sphere (calcium ions)

● = Cation coordination sphere (phosphate ions)

Figure 1.15. Schematic representation of biomineralization mechanisms in nature.

that of tissue regeneration achieved by either "in situ regeneration" or "tissue engineering" strategies [Hench & Polack 2002]. The former approach aims to use biomaterials that can be implanted into the damaged tissue to stimulate tissue regeneration through the activation of specific cell mechanisms. The second approach is based on the encapsulation of cells into porous biomaterial scaffolds to induce tissue regeneration either in a bioreactor system or upon implantation in the damaged tissue. In both cases, it is advocated that the biomaterial used will be able to mimic the components and signaling of natural tissues.

This book aims to explore the progress made in the field of biomaterials with biomimetic and/or bioactive properties and to assess the performances of those that have reached the clinical application stage. The particular scope of the next section of this chapter is to introduce the rationales underpinning biomimicry in the field of biomaterials.

1.4.1   Biomimicking of the ECM

Capitalizing on the performances of second-generation biomaterials and with a greater understanding of tissue regeneration processes, scientists have tried to synthesize novel biomaterials that mimic components of the ECM. In particular, polymers and metals have been functionalized with amino acid sequences involved in the cell–ECM recognition processes described in Section 1.3.1.1 (Figure 1.11). Biomaterials such as polyethylene glycol (PEG) hydrogels have been rendered bioactive through the inclusion of cell adhesion motifs (e.g., RGD) in their structure and bioresponsive by insertion of peptide sequences that are involved in enzyme-catalyzed reactions [Sakiyama-Elbert

& Hubbell 2001; Sanborn et al. 2002]. Substrates for Factor XIII have been exploited to induce the immediate cross-linking of hydrogels when in contact with blood [Sanborn et al. 2002]. Likewise, hydrogels have been synthesized that include peptidic substrates for the matrix metalloproteinase (MMP) enzyme secreted by tissue cells during their migration. This type of hydrogel is, therefore, able to respond to cell-released stimuli to change its structure and provide a route into the mesh for the migrating cell [Sakiyama-Elbert & Hubbell 2001]. Unlike second-generation biodegradable biomaterials (see Section 1.2.4), the MMP-functionalized hydrogels have transferred the control of bio-material degradation from spontaneous hydrolytic processes to biologically controlled events, thus tuning it to tissue regeneration.

Other biomimetic peptide sequences have been used to functionalize biomaterials and are listed in Table 1.2. These sequences are characterized by their affinity for specific cell types and their capacity to modulate specific cell functions (e.g., migration).

Although HA and Bioglass have set new standards in the development of biomaterials for bone regeneration, their mechanical properties significantly limit their surgical applications. These biomaterials are brittle, especially when porous, and are therefore difficult to adapt to the tissue defect during an operation. Also, when used as coatings, they tend to delaminate from the metal substrate, thus causing implant mobilization (Figure 1.16). In addition, the HA made available for clinical application has a degree of crystallinity significantly higher than bone apatite and, as a consequence, it cannot participate in the bone remodeling process (see Chapter 3). More recently, tricalcium phosphates have been introduced that can be used as bone cements, offering an alternative to the traditional PMMA-based materials. The relatively low degree of crystallinity makes these cements resorbable during bone remodeling, thus encouraging complete tissue regeneration. Progress has recently been made to make these biomaterials injectable and therefore suitable for different clinical applications where noninvasive surgery

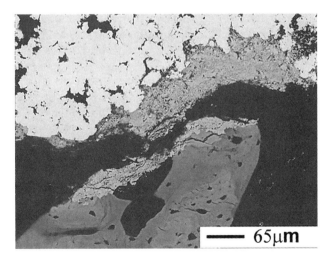

Figure 1.16. Ceramic coating delamination upon implantation. Micrograph shows bone integration with a delaminating coating.

is sought [Delgado et al. 2005]. Chapters 5 and 6 give an overview of the technological advances made in the attempt to mimic the texture of the ECM on the surface of metals and ceramics. Many investigations have been focused on establishing the optimal degree of surface roughness required to encourage cell adhesion. Anodization methods have been applied on titanium oxide surfaces, and optimal porosity has been sought in porous 3D ceramic scaffolds (see Chapters 5 and 6).

1.4.2   Biomimicking of Cell Membrane Components

More recently, biomimesis has been extended to the mimicking of the cell surface. The cell membrane (also called "plasmalemma") consists of a phospholipid bilayer that not only separates the cell from the extracellular space but also regulates interactions with the surrounding environment. These interactions are regulated by plasmalemma proteins such as the integrins (see Section 1.3.1.1), other ion channels, and glycoproteins that render the cell surface relatively hydrophilic (the glycocalyx). Chapter 4 illustrates typical examples of biomaterial surface functionalization that attempt to mimic these components. Polymers presenting phosphatidylcholine (PC) groups have been developed and used for different biomedical applications, while gels of PS have been obtained to induce surface mineralization of implants. Synthetic polymers such as poly(vinyl alcohol) and poly(2-hydroxyethyl methacrylate) have been modified by grafting sugar moieties such as dextran and galactose onto their surface in the attempt to simulate the cell glycocalyx (see Chapter 4).

1.4.3   Biomimicking Cell Signaling Pathways

Although fundamental to the control of tissue regeneration, cell-to-cell and cell-to-ECM recognition processes are complemented by an intricate network of autocrine and paracrine signaling (see Section 1.3.1.1). Cytokines and growth factors are secreted by immunocompetent and tissue cells during the different phases of healing to coordinate the activities required for the formation of new tissue [Martin & Leibovich 2005]. The efficacy of the signaling primarily depends on five factors:

1. Diffusion of the relevant bioactive molecules toward the target cell
2. Interaction with ECM components
3. Biorecognition of the relative cell receptor
4. Timeliness of the delivery
5. Composition of the bioactive molecule cocktail

Many of the third-generation biomaterials have tried to fulfill these goals by loading relevant growth factors in polymeric matrices, porous ceramics, and metal surfaces by entrapment or grafting [Hubbell 1999]. Such biomaterials are generally classified as bioactive biomaterials. Table 1.3 summarizes the growth factors commonly used in the field of biomaterials. Although promising results have been obtained (see Chapters 5, 6, and 7), none of the bioactive biomaterials so far engineered are able to fulfill all five criteria required to control tissue regeneration.

TABLE 1.3. Main Growth Factors Families Used in Bioactive Biomaterials

- Transforming growth factor superfamily
  - Transforming growth factor-β1
  - Bone morphogenetic proteins/osteopontins
- Growth differentiation factors
- Vascular endothelial growth factor
- Platelet-derived growth factor
- Fibroblast growth factor
- Epithelial growth factor
- Nerve growth factor

Thus far, biomaterials loaded with growth factors have tried to fulfill only the requirements of points 1 and 2 since they have been optimized to deliver specific growth factors in a controlled manner. This delivery is not tuned with the expression of the relevant receptors on the cell surface, which depends on cell phenotype and tissue regeneration phase (points 3 and 4). Moreover, most of the bioactive biomaterials tested at the research and clinical level focus on the delivery of only one type of growth factor, thus neglecting the importance of combined release, which leads to a physiological tissue regeneration (point 5).

In general, the optimization of growth factor delivery has been pursued by following traditional drug delivery approaches: The bioactive components are loaded into the biomaterial carrier or grafted on its surface through chemical bonding that is prone to cleavage in biological media. Kinetic release studies are usually performed by traditional pharmacological methods and bioactivity studied by in vitro cell experiments or by animal models. In this manner, the delivery of growth factors to promote cell proliferation (TGF-β1), osteoblast differentiation (BMP-2), vascularization (angiogenic factors), and nerve regeneration (neuronal survival and differentiation factors) has been optimized in different biomaterial carriers [Hubbell 1999]. The ability of growth factors to bind components of the ECM has been exploited by materials scientists to control the delivery of the bioactive molecules to cells resident within tissue engineering constructs or in the surrounding tissue. For example, collagen has been functionalized with heparin to utilize the ability of TGF-β1 to bind to heparin domains [Schroeder-Tefft et al. 1997].

1.4.3.1 Modulation of the Growth Factor Signaling by Gene Expression: Bioactive Gene Delivery Systems. In an alternative approach, bioactivity of biomaterials has been pursued by gene-delivery strategies. Plasmid DNA presented to cells by a biomaterial has been successfully employed to enhance the gene expression and synthesis of growth factors important to tissue regeneration. This approach has been applied to suture materials as well as to implants to enhance the regeneration of cardiovascular and bony tissues [Hubbell 1999]. Synthetic biomimetic polymers have also been designed that present peptides to the cells that are usually exposed on viral coats, thus facilitating their penetration into the cell by endocytosis [Hoffman et al. 2002]. These biomimetic biomaterials have been loaded with plasmid DNA grafted

onto the synthetic polymers via bonds that are sensitive to the relatively low pH of the endosomes, the vesicles used by the cell to engulf material from the extracellular space. The cleavage of the bonding in the intracellular space thus leads to the release of the plasmid DNA. The benefit of the gene-delivery approach is the prolonged effect on the synthesis of growth factor and therefore, its sustained production throughout the tissue regeneration process.

1.5   BIOACTIVE BIOMATERIALS FROM DIFFERENT NATURAL SOURCES

Nature has evolved common mechanisms to regulate biological processes. Biomolecular recognition, cell interactions, mineralization and, more broadly, bioactivity, share similar pathways across organisms, which have triggered the development of new bioactive biomaterials. Beyond the poorly clarified hemostatic properties of alginate, new biomaterials have recently emerged that appear to be able to participate in the tissue regeneration processes in an active manner. In this section, two main examples are given: silk fibroin- and soybean-based biomaterials.

1.5.1   Silk Fibroin

Silk is secreted from the silkworm gland as fibers composed by an inner core made of a protein called fibroin, which is coated by a second protein called sericin (Figure 1.17). The excellent mechanical properties of silk have been long been exploited to produce suturing material. However, in the 1980s, some investigations highlighted the adverse reaction elicited by virgin silk when in contact with healing ocular tissues. Conversely,

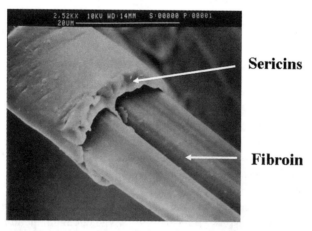

Figure 1.17. Scanning electron micrograph of a typical silkworm silk fiber showing the fibroin bavellae emerging from the sericin coating. (Photo by Dr. A Motta, University of Trento, Italy).

it was noticed that when silk sutures deprived of the sericin coating were used, these adverse reaction no longer occurred [Soong & Kenyon 1984]. The ability to remove the sericin coat from virgin silk, as well as the possibility of engineering fibroin in the form of films and membranes, triggered interest in developing fibroin-based biomaterials. However, no clear assessment of the immunological properties of this material was made until recently when Santin et al. (1999) proved that engineered fibroin films elicited a significantly lower immune response in vitro than other conventional polymers. Later studies confirmed these early findings and offered more insights about the degree of interactions that silk fibroin can establish with the host environment depending on its native or denatured conformation [Motta et al. 2002; Santin et al. 2002; Meinel et al. 2005]. These investigations showed that when in fiber form, silk fibroin can elicit the activation of inflammatory cells and exhibit procoagulant activity by interacting with the polymerizing fibrin (see Section 1.3.1). Beyond the low immunogenicity, denatured fibroin biomaterials have been shown to stimulate cell proliferation in vitro and support bone regeneration in rabbit models [Motta et al. 2004; Fini et al. 2005]. Although it has been postulated that the cell proliferation stimulus may be derived from specific peptides present in the fibroin structure, no sequence has yet been identified to prove this hypothesis. It is not clear whether the different levels of biointeraction are elicited by the transition from the β-sheet structure of the fiber to the amorphous conformation of the denatured protein films. These conformational changes may lead either to different surface physicochemical properties or to an altered exposure of specific bioligands.

1.5.2   Soybean-Based Biomaterials

Soybean is one of the most commonly used foods throughout the world. The bean is composed of proteins (40%), carbohydrates (38%), lipids (18%), and minerals (4%) (Figure 1.18). Plant estrogens called isoflavones are also included, which are present in both a glycosylated and nonglycosylated form [Murkies et al. 1998]. The glycosylated forms genistin and daidzin (Figure 1.19a,b) are the most abundant forms and are accompanied by the nonglycosylated genistein and daidzein (Figure 1.19c,d). Glycosylated isoflavones can be transformed by spontaneous- or enzyme-driven hydrolysis into the respective nonglycosylated forms [Walle et al. 2005]. It has been proven that the nonglycosylated form can exert different effects on eukaryotic cells [Middleton et al. 2000]. In general, they can reduce the activation of immunocompetent cells and tissue cell proliferation. Isoflavones can inhibit tyrosine kinase receptors on the immunocompetent cell plasmalemma or act as anti-redox compounds able to buffer free radicals produced by inflammatory cells. It has also been proven that isoflavones can penetrate the cell membrane and interact with the estrogen receptor β of the nuclear membrane. The binding of isoflavones to this receptor leads to an inhibition of the cell cycle in its G1/M phase, thus reducing the cell proliferation rate. Simultaneously, the isoflavones are responsible for an increased cell differentiation, which leads to an induced collagen synthesis both in vitro and in vivo. In the case of osteoblasts, isoflavones seem to stimulate the differentiation of the cells by inducing the synthesis of ALP and BMP-2 [Zhou et al. 2003; Morris et al. 2006]. Indeed, it is believed that the low incidence of

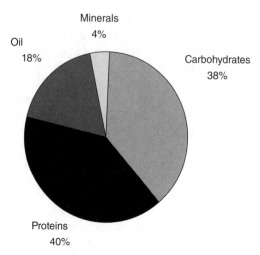

Figure 1.18. Soybean composition. A novel class of biomaterials has recently been developed by thermosetting the defatted soybean curd (Santin et al. 2001).

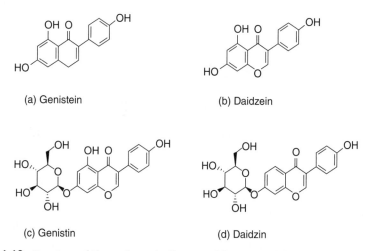

(a) Genistein

(b) Daidzein

(c) Genistin

(d) Daidzin

Figure 1.19. Structure of the main soy isoflavones: (a) genistin, (b) daidzin, (c) genistein, (d) daidzein.

both breast and prostate cancers, and osteoporosis in populations with a soy-rich diet, can be attributed to the isoflavone bioactivity [Middleton et al. 2000].

Soybeans have long been proposed as sources of biodegradable materials for many applications. Henry Ford invested money in research to make car body parts from soy. Glues, fibers, and other objects can be fabricated from the soybean fraction. Soy protein has also been used to make biodegradable biomaterials.

Figure 1.20. Soybean-based biodegradable membrane for surgical application.

More recently, a novel class of biodegradable biomaterials has been introduced that is produced by thermosetting defatted soy curd [Santin et al. 2001] (Figure 1.20). The advantage of this class of soy-based biomaterials is the preservation of the protein, carbohydrate, and isoflavone components that are able to modulate the inflammatory response and tissue regeneration while degrading. These biomaterials are easy to handle during surgical procedures, and animal models have been shown to favor the regeneration of both soft and bony tissues while ruling out the formation of a fibrotic capsule. Histocytochemical analysis seems to suggest that soy's ability to regenerate tissue without forming fibroses is due to its maintenance of the inflammatory response in its acute phase rather than triggering chronic inflammation. The tissues surrounding the soy implants are indeed characterized by a prevalence of neutrophils rather than monocytes/macrophages (see Section 1.3.1). The isoflavone-driven induction of collagen synthesis may also be a key factor in the achievement of tissue regeneration at the implantation site.

Future insights into the biological pathways driven by soybean-based biomaterials may lead to the development of new biomaterials capable of simultaneously modulating both biochemical signaling and ECM components.

1.6 SCOPE OF THIS BOOK

This book is aimed at professionals already working in the field of biomedical materials and as a text suitable for students at the graduate and postgraduate levels. In the following chapters we aim to offer a comprehensive overview that highlights the effects of different types of biomaterials in the context of tissue regeneration and/or tissue functionality. Specifically, the links between tissue functionality and model clinical applications to chemical and engineering solutions will be demonstrated. Each chapter includes problem sets and questions to allow revision of the different topics covered.

REFERENCES

Anderson JM (1988) Inflammatory response to implants. ASAIO 11: 101–107.

Anderson JM (2001) Biological responses to materials. Annual Review of Materials Research 31: 81–110.

Armulik A, Velling T, Johansson S (2004) The integrin β1 subunit transmembrane domain regulates phosphatidylinositol 3-kinase-dependent tyrosine phosphorylation of Crk-associated substrate. Molecular Biology of the Cell 15: 2558–2567.

Asano Y, Ihn H, Yamane K, Jinnin M, Tamaki K (2006) Increased expression of integrin alpha v beta 5 induces the myofibroblastic differentiation of dermal fibroblasts. American Journal of Pathology 168: 499–510.

Bach AD, Bannasch H, Galla TJ, Bittner KM, Stark GB (2001) Fibrin glue as matrix for cultured autologous urothelial cells in urethral reconstruction. Tissue Engineering 7(1): 45–53.

Barbucci R, Magnani A, Bazkin A, Dacosta ML, Bauser H, Hellwig G, Martuscelli E, Cimmino S (1993) Physicochemical surface characterization of hyaluronic-acid derivatives as a new class of biomaterials. Journal of Biomaterials Science. Polymer Edition 4: 245–273.

Boccafoschi F (2005) Biological performances of collagen-based scaffolds for vascular tissue engineering. Biomaterials 26: 7410.

Boskey AL (1978) Role of calcium-phospholipid-phosphate complexes in tissue mineralization. Metabolic Bone Disease & Related Research 1: 137–142.

Boulter E, van Obberghen-Schilling E (2006) Integrin-linked kinase and its partners: a modular platform regulating cell-matrix adhesion dynamics and cytoskeletal organization. European Journal of Cell Biology 85: 255–263.

Buckwalter JA, Glimcher MJ, Cooper RR, Recker R (1995) Bone biology: Part I. Structure, blood supply, cells, matrix, and mineralization. The Journal of Bone and Joint Surgery 77-A: 1256–1289.

Buhling F, Groneberg D, Welte T (2006) Proteases and their role in chronic inflammatory lung diseases. Current Drug Targets 7: 751–759.

Cabodi S, Moro L, Bergatto E, Boeri Erba E, Di Stefano P, Turco E, Tarone G, De Filippi P (2004) Integrin regulation of epidermal growth factor (EGF) receptor and of EGF-dependent responses. Biochemical Society Transactions 32: 438–442.

Camacho NP, Rinnerthaler S, Paschalis EP, Mendelsohn R, Boskey AL, Fratzl P (1999) Complementary information on bone ultrastructure from scanning small angle X-ray scattering and Fourier-transform infrared microspectroscopy. Bone 25: 287–293.

Carreño-Gómez B, Duncan R (1997) Evaluation of the biological properties of soluble chitosan and chitosan microspheres. International Journal of Pharmaceutics 148(2): 231–240.

Charnley J (1961) Arthroplasty of the hip—a new operation. Lancet 1: 1129–1123.

Clarke HJ, Jinnah RH, Lennox D (1990) Osteointegration of bone-graft in porous-coated total hip arthroplasty. Clinical Orthopaedics and Related Research 258: 160–167.

Day AJ, Prestwich GD (2002) Hyaluronan-binding proteins: tying up the giant. The Journal of Biological Chemistry 277: 4585–4588.

Delgado JA, Harr I, Almirall A, del Valle S, Planell JA, Ginebra MP (2005) Injectability of a macroporous calcium phosphate cement. Key Engineering Materials 284-286: 157–160.

Engel J, Bachinger HP (2005) Structure, stability and folding of the collagen triple helix. Collagen Topics in Current Chemistry 247: 7–33.

Fincham AG, Simmer JP (1997) Amelogenin proteins of developing dental enamel. In: Dental enamel. Ciba Foundation Symposium 118–134.

Fini M, Motta A, Torricelli P, Glavaresi G, Aldini NN, Tschon M, Giardino R, Migliaresi C (2005) The healing of confined critical size cancellous defects in the presence of silk fibroin hydrogel. Biomaterials 26: 3527–3536.

Fraser JRE, Laurent TC, Laurent UBG (1997) Hyaluronan: its nature, distribution, functions and turnover. Journal of Internal Medicine 242: 27–33.

Fukushi J, Makagiansar IT, Stallcup WB (2004) NG2 proteoglycan promotes endothelial cell motility and angiogenesis via engagement of galectin-3 and $\alpha 3\beta 1$ integrin. Molecular Biology of the Cell 15: 3580–3590.

Gentleman E, Lay AN, Dickerson DA, Nauman EA, Livesay GA, Dee KC (2003) Mechanical characterisation of collagen fibres and scaffolds for tissue engineering. Biomaterials 24: 3805–3813.

Gentleman E, Livesay GA, Dee KC, Nauman EA (2006) Development of ligament-like structural organization and properties in cell-seeded collagen scaffolds in vitro. Annals of Biomedical Engineering 34(5): 726–736.

Giusti P, Lazzeri L, Cascone MG, Seggiani M (1995) Blends of synthetic and natural polymers as special performance materials. Macromolecular Symposia 100: 81–87.

Goessler UR, Bieback K, Bugert P, Heller T, Sadick H, Hormann K, Riedel F (2006) In vitro analysis of integrin expression during chondrogenic differentiation of mesenchymal stem cells and chondrocytes upon dedifferentiation in cell culture. International Journal of Molecular Medicine 17: 301–307.

Goldrin K, Jones GE, Thiagarajah R, Watt DJ (2001) The effect of galectin-1 on the differentiation of fibroblasts and myoblasts in vitro. Journal of Cell Science 115: 355–366.

Grayson ACR, Voskerician G, Lynn A, Anderson JM, Cima MJ, Langer R (2004) Differential degradation rates in vivo and in vitro of biocompatible poly(lactic acid) and poly(glycolic acid) homo- and co-polymers for a polymeric drug-delivery microchip. Journal of Biomaterials Science. Polymer Edition 15: 1281–1304.

Hayashi T (1994) Biodegradable polymers for biomedical uses. Progress in Polymer Science 9: 663–702.

Helgerson SL, Seelich T, DiOrio JP, Tawil B, Bittner K, Spaethe R (2004) Fibrin. In: Encyclopaedia of Biomedical Engineering, 603–610. New York: Marcel Dekker Inc.

Hench LL, Polack J (2002) Third-generation biomedical materials. Science 295: 1014–1017.

Hing KA, Best SM, Tanner KE, Bonfield W, Ravell PA (2004) Mediation of bone ingrowth in porous hydroxyapatite bone graft substitutes. Journal of Biomedical Materials Research. Part A 68A: 187–200.

Hoffman AS, Stayton PS, Press O, Murthy N, Lackey CA, Cheung C, Black F, Campbell J, Fausto N, Kyriakides TR, Bornstein P (2002) Design of "smart" polymers that can direct intracellular drug delivery. Polymers for Advanced Technologies 13: 992–999.

Hubbell JA (1999) Bioactive biomaterials. Current Opinion in Biotechnology 10: 123–129.

Jayakumar R, Reis RL, Mano JF (2006) Phosphorous containing chitosan beads for controlled oral drug delivery. Journal of Bioactive and Biocompatible Polymers 21(4): 327–340.

Jones V, Grey JE, Harding KG (2006) Wound dressings. British Medical Journal 332: 777–780.

Juliano RL, Redding P, Alahari S, Edin M, Howe A, Aplin A (2004) Integrin regulation of cell signalling and motility. Biochemical Society Transactions 32: 443–446.

Kolff WJ (1993) The beginning of the artificial kidney. Artificial Organs 17(5): 293–299.

Kutschka I, Chen IY, Kofidis T, Arai T, von Degenfeld G, Sheikh AY, Hendry SL, Pearl J, Hoyt G, Sista R, Yang PC, Blau HM, Gambhir SS, Robbins RC (2006) Collagen matrices enhance survival of transplanted cardiomyoblasts and contribute to functional improvement of ischemic rat hearts. Circulation 114(Suppl. 1): I167–I173.

Li R, Bennett JS, DeGrado WF (2004) Structural basis for integrin αIIβ$_3$ clustering. Biochemical Society Transactions 32: 412–415.

Madihally SV, Matthew HWT (1999) Porous chitosan scaffolds for tissue engineering. Biomaterials 20: 1133–1142.

Mahanthappa NK, Cooper DN, Barondes SH, Schwarting GA (1994) Rat olfactory neurons can utilize the endogenous lectin, L-14, in a novel adhesion mechanism. Development 120: 1373–1384.

Mann S (1988) Molecular recognition in biomineralization. Nature 332: 119–124.

Martin P, Leibovich SJ (2005) Inflammatory cells during wound, repair: the good, the bad and the ugly. Trends in Cell Biology 15: 599–607.

Meinel L, Hofmann S, Karageorgiou V, Kirker-Head C, McCool J, Gronowicz G, Zichner L, Langer R, Vunjak-Novakovic G, Kaplan DL (2005) The inflammatory responses to silk films in vitro and in vivo. Biomaterials 26: 147–155.

Middleton E, Kandaswami C, Theoharides TC (2000) The effects of plant flavonoids on mammalian cells: implication for inflammation, heart disease, and cancer. Pharmacological Reviews 52: 673–751.

Minuth WW, Sittinger M, Kloth S (1998) Tissue engineering: generation of differentiated artificial tissues for biomedical applications. Cell and Tissue Research 291: 1–11.

Morris C, Thorpe J, Ambrosio L, Santin M (2006) The soybean isoflavone genistein induces differentiation of MG63 human osteosarcoma osteoblasts. The Journal of Nutrition 136: 1166–1170.

Motta A, Migliaresi C, Lloyd AW, Denyer SP, Santin M (2002) Serum protein absorption on silk fibroin fibers and films: surface opsonization and binding strength. Journal of Bioactive and Compatible Polymers 17: 23–35.

Motta A, Migliaresi C, Faccioni F, Torricelli P, Fini M, Giardino R (2004) Fibroin hydrogels for biomedical applications: preparation, characterization and in vitro cell culture studies. Journal of Biomaterials Science. Polymer Edition 15: 851–864.

Murkies AL, Wilcox G, Davies SR (1998) Phytoestrogens. The Journal of Clinical Endocrinology and Metabolism 83: 297–303.

Mytar B, Diedlar M, Woloszyn M, Colizzi V, Zembala M (2001) Cross-talk between human monocytes and cancer cells during reactive oxygen intermediates generation: the essential role of hyaluronan. International Journal of Cancer 94: 727–732.

Norde W (1986) Adsorption of proteins from solution at the solid-liquid interface. Advances in Colloid and Interface Science 25: 267–340.

Phillips JM, Kao WJ (2005) Macrophage adhesion on gelatine-based interpenetrating networks grafted with PEGylated RGD. Tissue Engineering 11: 964–973.

Raghunath K, Rao KP, Joseph KT (1980) Chemically modified natural polymers as biomaterials: 1. Polysaccharide-gelatin conjugates. Polymer Bulletin 2: 477–483.

Rao SB, Sharma CP (1997) Use of chitosan as a biomaterial: studies on its safety and hemostatic potential. Journal of Biomedical Materials Research 34: 21–28.

Rezania A, Healy KE (1999) Biomimetic peptide surfaces that regulate adhesion, spreading, cytoskeletal organization, and mineralization of the matrix deposited by osteoblast-like cells. Biotechnology Progress 15: 19–32.

Rivier N, Sadoc JF (2006) The transverse structure of collagen. Philosophical Magazine 86(6–8): 1075–1083.

Rosca C, Popa MI, Chitanu GC, Popa M (2005) Interaction of chitosan with natural or synthetic anionic polyelectrolytes: II. The chitosan-alginate complex. Cellulose Chemistry and Technology 39(5–6): 415–422.

Rothenburger M (2001) In vitro modelling of tissue using isolated vascular cells on a synthetic collagen matrix as a substitute for heart valves. The Thoracic and Cardiovascular Surgeon 49: 204.

Roughley P, Hoemann C, DesRosiers E, Mwale F, Antoniou J, Alini M (2006) The potential of chitosan-based gels containing intervertebral disc cells for nucleus pulposus supplementation. Biomaterials 27: 388–396.

Rücker M, Laschke MW, Junker D, Carvalho C, Schramm A, Mülhaupt R, Gellrich N, Menger MD (2006) Angiogenic and inflammatory response to biodegradable scaffolds in dorsal skinfold chambers of mice. Biomaterials 27(29): 5027–5038.

Sakiyama-Elbert SE, Hubbell JA (2001) Functional biomaterials: design of novel biomaterials. Annual Review of Materials Research 31: 183–201.

Sanborn TJ, Messersmith PB, Barron AE (2002) In situ crosslinking of a biomimetic peptide-PEG hyrogel via thermally triggered activation of factor XIII. Biomaterials 23: 2703–2710.

Sandrini E, Morris C, Chiesa R, Cigada A, Santin M (2005) In vitro assessment of the osteointegrative potential of a novel multiphase anodic spark deposition coating for orthopaedic and dental implants. Journal of Biomedical Materials Research Part B: Applied Biomaterials 73(2): 392–399.

Santin M, Cannas M (1999) Collagen-bound alpha(1)-microglobulin in normal and healed tissues and its effects on immunocompetent cells. Scandinavian Journal of Immunology 50: 289–295.

Santin M, Wassall MA, Peluso G, Denyer SP (1997) Adsorption of α-1-microglobulin from biological fluids onto polymer surfaces. Biomaterials 18: 823–827.

Santin M, Motta A, Freddi G, Cannas M (1999) In vitro evaluation of the inflammatory potential of the silk fibroin. Journal of Biomedical Materials Research 46: 382–389.

Santin M, Nicolais L, Ambrosio L (2001) Soybean-based biomaterials PCT/GB01/03464.

Santin M, Lloyd AW, Denyer SP, Motta A (2002) Domain-driven binding of fibrin(ogen) to silk fibroin. Journal of Bioactive and Compatible Polymers 17: 195–208.

Schroeder-Tefft JA, Bentz H, Estridge TD (1997) Collagen and heparin matrices for growth factor delivery. Journal of Controlled Release 49: 291–298.

Sharon N, Lis H (1993) Carbohydrates in cell recognition. Scientific American January Issue: 73–81.

Soong HK, Kenyon KR (1984) Adverse reactions to virgin silk sutures in cataract surgery. Ophthalmology 91: 479–483.

Stefansson A, Armulik A, Nilsson IM, von Heijne G, Johansson S (2004) Determination of N- and C-terminal borders of the transmembrane domain of integrin subunits. The Journal of Biological Chemistry 279: 21200–21205.

Steflik DE, Corpe RS, Lake FT, Young RT, Sisk AL, Parr GR, Hanes PJ, Berkery DJ (1998) Ultrastructural analyses of the attachement (bonding) zone between bone and implanted biomaterials. Journal of Biomedical Materials Research 39: 611–620.

Takagi J (2004) Structural basis for ligand recognition by RGD (Arg-Gly-Asp)-dependent integrins. Biochemical Society Transactions 32: 403–406.

Takemoto M, Fuhibayashi S, Neo M, Suzukin J, Kokubo T, Nakamura T (2005) Mechanical properties and osteoconductivity of porous bioactive titanium. Biomaterials 26: 6014–6023.

Takeshita F, Ihama S, Ayukawa Y, Akedo H, Suetsugu T (1997) Study of bone formation around dense hydroxyapatite implants using light microscopy, image processing and confocal laser scanning microscopy. Biomaterials 18: 317–322.

Teder P, Vandivier RW, Jiang DH, Liang JR, Cohn L, Pure E, Henson PM, Noble PW (2002) Resolution of lung inflammation by CD44. Science 296: 155–158.

Tomihata K, Ikada Y (1997) In vitro and in vivo degradation of films of chitin and its deacetylated derivatives. Biomaterials 18: 567–575.

Turley EA, Noble PW, Bourguignon LYW (2002) Signalling properties of hyaluronan receptors. The Journal of Biological Chemistry 277: 4589–4592.

Turley EA (2001) Extracellular matrix remodelling: multiple paradigms in vascular disease. Circulation Research 88: 2–4.

Velling T, Risteli J, Wennerberg K, Mosher DF, Johansson S (2002) Polymerization of type I and III collagens is dependent on fibronectin and enhanced by integrins $\alpha_{11}\beta_1$ and $\alpha_2\beta_1$. The Journal of Biological Chemistry 277: 37377–37381.

Walle T, Browning AM, Steed LL, Reed SG, Walle UK (2005) Flavonoids glucosides are hydrolyzed and thus activated in the oral cavity in humans. The Journal of Nutrition 135: 48–52.

Wells V, Malluchi L (1991) Identification of an autocrine negative growth factor: murine beta-galactoside-binding protein is a cytostatic factor and cell regulator. Cell 64: 91–97.

Williams DF, ed. (1987) Definitions in biomaterials. In: Proceedings of a Consensus Conference of the European Society for Biomaterials, Chester, England, 3–5 March 1986, Progress in Biomedical Engineering Vol. 4, Elsevier, Amsterdam.

Williams HP (2001) Sir Harold Ridley's Vision. The British Journal of Ophthalmology 85: 1022–1023.

Xynos ID, Edgar AJ, Buttery LDK, Hench LL, Polak JM (2000) Ionic products of bioactive glass dissolution increase proliferation of human osteoblasts and induce insulin-like growth factor II mRNA expression and protein synthesis. Biochemical and Biophysical Research Communications 276: 461–465.

Zaidel-Bar R, Cohen M, Addadi L, Geiger B (2004) Hierarchical assembly of cell-matrix adhesion complexes. Biochemical Society Transactions 32: 416–420.

Zhai Y, Cui FZ (2006) Recombinant human-like collagen directed growth of hydroxyapatite neocrystals. Journal of Crystal Growth 291: 202–206.

Zhou S, Turgeman G, Harris SE, Leitman DC, Komm BS, Bodine PVN, Gazit D (2003) Estrogens activate bone morphogenetic protein-2 gene transcription in mouse mesenchymal stem cells. Molecular Endocrinology 17: 56–66.

2

SOFT TISSUE STRUCTURE AND FUNCTIONALITY

Gabriela Voskerician

2.1 OVERVIEW

A tissue is an aggregate or a group of cells that functions in a collective manner to perform one or more specific physiological tasks. Even though organs are characterized by a large diversity of structures and functions, the specific tissues present can be classified into only four basic categories: *epithelium, connective tissue, muscle tissue,* and *nerve tissue.*

The epithelium lines the cavities and surfaces of structures throughout the body and forms glands. The connective tissue is a category of exclusion rather than precise definition as it represents everything else outside of the other three basic tissue types. The generic role of the connective tissue is to support, structurally and functionally, the other three tissue types. The muscle tissue consists of cells which have the ability to contract and to conduct electrical impulses, thus being responsible for movement. The nerve tissue receives, transmits, and integrates information from outside and inside the body to control and coordinate the communication between different parts of the body, as well as between different parts of the body and the surrounding environment. The treatment of the nervous tissue has not been undertaken here due to the complexity of the subject which deserves its own independent chapter, as well as in an attempt to uphold the subject treatment adopted by the following chapters. Finally, the structure and function of bone, a hard connective tissue, has been extensively captured in Chapter 3 of this book.

Biomimetic, Bioresponsive, and Bioactive Materials: An Introduction to Integrating Materials with Tissues, First Edition. Edited by Matteo Santin and Gary Phillips.
© 2012 John Wiley & Sons, Inc. Published 2012 by John Wiley & Sons, Inc.

The presence of a foreign body, a biomaterial for example, leads to an associated inflammatory and wound healing response during which the body responds not only to the infliction of the mechanical injury (the incision/tissue manipulation required to place the biomaterial) but also to the presence of the foreign body. The development of biomaterials in the context of tissue regeneration and/or tissue functionality must rely on superior understanding of the body's responses to injury and its approach to healing. The field of biomaterials has entered a new age in which biomaterial design is aimed toward tissue regeneration (restoration of native structure and function) rather than repair (fibrous encapsulation), a paramount distinction which must be emphasized. Consequently, a brief treatment of the subject of inflammation and wound healing concludes this chapter, as a reminder of the challenges which stand before us.

A word of gratitude is owed to my mentor and collaborator, James M. Anderson, MD, PhD, Department of Pathology, Case Western Reserve University, Cleveland, and my esteemed colleagues, Lutz Slomianka, PhD, Department of Anatomy, University of Zürich, Zürich, Switzerland and Ed Friedlander, MD, Department of Medicine and Biosciences, Kansas City University, Kansas City, for providing me with many of the histological images featured in this chapter. For uniformity within the histological material presented, drawn from various sources, no image magnifications are included.

2.2 EPITHELIAL TISSUE

2.2.1 Background

The epithelia are a diverse group of avascular sheet-like tissues which, with rare exceptions, cover or line all body surfaces, cavities, and body tubes that communicate with the exterior. Under specific circumstances, the epithelia also form the secretory portion of glands and their ducts. Thus, the epithelia function as interfaces between different biological compartments mediating a wide range of activities such as selective diffusion, absorbtion and/or secretion, as well as physical protection and containment [Ross et al. 2003]. The epithelial cells have four principal characteristics:

- They form cell junctions through close apposition and adherence to one another by means of specific cell-to-cell adhesion molecules.
- They exhibit three distinct morphological domains: apical domain (free surface), lateral domain, and basal domain responsible for morphological polarity through specific membrane proteins.
- Their basal surface is attached to an underlying basement membrane, a noncellular protein-polysaccharide rich layer.
- They are dependent on the diffusion of oxygen and metabolites from the adjacent supporting tissue.

In specific situations, the epithelial cells lack an apical surface, such as in the case of epitheloid tissue (e.g., endocrine glands) [Young 2000].

The epithelia are traditionally classified, taking into consideration two factors: the number of cell layers and the shape of the surface cells. Thus, the terminology reflects structure rather than function [Ross et al. 2003] (Tables 2.1 and 2.2). Consequently, the epithelia are described as

- *Simple*, when a single layer of epithelial cells is present,
- *Stratified*, when two or more layers of epithelial cells are present.

Further, the individual cells composing the epithelium are described as

- *Squamous*, when the width of the cell is greater than its height,
- *Cuboidal*, when the width, depth, and height are approximately the same,
- *Columnar*, when the height of the cell appreciably exceeds the width.

Additions to the above classification format include the pseudostratified and transitional epithelia:

- *Pseudostratified*, a simple epithelium resting on the basement membrane where the epithelium appears to be stratified due to some of the cells not reaching the free surface.
- *Transitional*, it is applied to the epithelium lining the lower urinary tract presenting specific morphological characteristics that allow it to distend.

The histological images featured in Tables 2.1 and 2.2 were obtained by subjecting the tissue samples to hematoxylin and eosin (H&E) staining. This nonspecific tissue stain is used to primarily display general structural features of a tissue. The organic solvents required for the H&E tissue processing preclude the preservation within the tissue sample of particular components such as any soluble complexes (e.g., neutral lipids) and small molecules (e.g., glucose).

2.3   THE SKIN

The skin forms the external covering of the body representing its largest organ and being responsible for 15–20% of its total mass. Severely compromising the skin integrity through injury or illness leads to enhanced physiological imbalance and ultimately to disability or even death.

2.3.1   Structure and Functionality

The skin and its derivatives (hair, nails, and sweat glands) constitute the integumentary system. The skin has four major functions [Young 2000]:

- *Protection*, provides protection against ultraviolet light as well as mechanical, chemical, and thermal insults; in addition, its relatively impermeable surface

TABLE 2.1. Epithelial Tissue Types: Simple (Histological Images Courtesy of Ed Friedlander, MD)

Classification and Function[a]	Diagram	Image
Simple squamous: Lines surfaces involved in passive transport of either gases (lungs) or fluids (walls of blood vessels, featured, H&E)		
Simple cuboidal: Lines small ducts and tubules which may have excretory, secretory, or absorbtive function (pancreatic ducts, featured, H&E)		
Simple columnar: Lines highly absorbtive surfaces (small intestine) as well as highly secretory surfaces (stomach, featured, H&E)		
Pseudostratified columnar: Exclusively lines the large airways of the respiratory system (large airways, featured, H&E)		

[a] The identified functions and locations represent some typical examples.

TABLE 2.2. Epithelial Tissue Types: Stratified (Histological Images Courtesy of Ed Friedlander, MD)

Classification and Function[a]	Diagram	Image
Stratified squamous: Lines highly abrasive surfaces being instrumental in protecting underlying tissues (skin, featured, H&E)		
Stratified cuboidal: Lines the larger excretory ducts of the exocrine glands (sweat gland, featured, H&E)		
Transitional: Almost exclusively lines the urinary tract (urinary tract, featured, H&E)		

[a] The identified functions and locations represent some typical examples.

prevents rapid dehydration and acts as a physical barrier to microorganism attack;

• *Sensation*, as the largest sensory organ in the body, the skin contains a variety of receptors for touch, pressure, pain, and temperature;

• *Thermoregulation*, the insulation against heat loss through the presence of surface pilosity and subcutaneous adipose tissue; the heat loss is facilitated by evaporation of sweat from the skin surface and increased blood flow through available vascular networks;

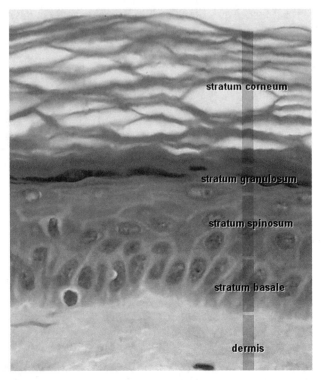

Figure 2.1. Epidermis (H&E). The layers of the epidermis are shown in this section of thin skin. The mitotically active cells of the stratum basale mature as they migrate through the epidermis layers toward the free surface of the skin. The rate of mitosis generally equals the rate of desquamation of the keratin present in stratum corneum (courtesy of Lutz Slomianka, PhD).

- *Metabolic function*, as a result of the subcutaneous adipose tissue storage of energy in the form of triglycerides, in addition to the Vitamin D synthesis which supplements that derived from dietary sources.

The layers of the skin include the *epidermis*, *dermis*, and *hypodermis*.

The *epidermis* is composed of stratified squamous epithelium featuring four distinct layers, while thick skin is composed of five layers [Young 2000] (Figure 2.1):

- *Stratum basale*, also known as *stratum germinativum* because of the presence of mitotically active cells, the stem cells of the epidermis providing a constant supply of keratinocytes to replace those lost;
- *Stratum spinosum*, termed as spinous because of the light microscopy appearance of short processes extending from cell to cell; the cells of this layer are in the process of growth and early keratin synthesis;

- *Stratum granulosum*, contains numerous intensely staining granules which contribute to the process of keratinization;
- *Stratum lucidum*, is limited to thick skin and is considered a subdivision of the stratum corneum;
- *Stratum corneum*, composed of keratinized cells.

The epithelial classification of the stratum basale is simple cuboidal or low columnar separated from the dermis by a basement membrane. The basal aspect of the germinative cells (stem cells) is irregular and bound to the basement membrane by numerous anchoring filaments (hemidesmosomes). The cells of the stratum spinosum are relatively large and polyhedral in shape. Prominent nucleoli and cytoplasmic basophilic granules indicate active protein synthesis. Cytokeratin is the primary protein synthesized by the cells of stratum spinosum. The cells of stratum granulosum are characterized by basophilic granules that populate most of the cytoplasm. In the outmost layer of stratum granulosum, cell death occurs as a result of lysosomal membrane rupture. Released lysosomal enzymes seem to be involved in the process of keratinization. The cells of the stratum corneum are dead or in the process of dying. They appear flattened, devoid of nuclei and other organelles, and filled with mature keratin [Young 2000].

The junction between the epidermis and dermis appears as an uneven boundary. The dermal layer of the skin provides a flexible but robust base for the epidermis. It also contains an extensive vascular supply for the metabolic support of the avascular epidermis as well as for thermoregulation [Young 2000]. The dermis is divided into two layers (Figure 2.2):

- *papillary dermis*, finger-like connective tissue protrusions that project into the undersurface of the epidermis,
- *reticular dermis*, composed of irregular bundles of collagen within which blood vessels are present joining the plexus of vessels in the papillary dermis with the larger, deeper vessels at the junction between dermis and subcutis.

Elastin is a significant component of both dermal layers (Figure 2.3). The elastin and collagen fibers present in the dermis are illustrated in Figure 2.3 through the use of the Verhoeff–van Gieson stain. The Verhoeff component of the stain is specific to the elastin fibers (stain gray/black), while the van Gieson component is specific for collagen fibers (stain red). The staining of this tissue sample with H&E would preclude color-specific visualization of the elastin and collagen fibers. In the reticular layer, the elastin fibers are robust and follow the direction of the collagen bundles, creating regular lines of tension in the skin, called Langer's lines [Ross et al. 2003]. In the papillary zone, the elastin fibers are thin and significantly less dense. The cellular component of the dermis is mainly the fibroblast, responsible for the production of collagen and elastin. In addition, small populations of lymphocytes, mast cells, and tissue macrophages are present for the purpose of nonspecific defense and immune surveillance.

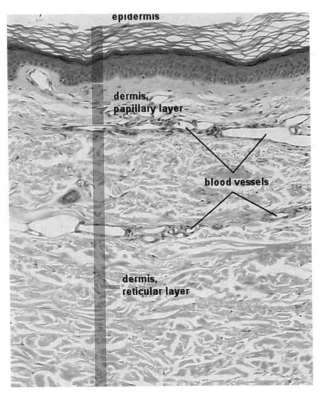

Figure 2.2. Dermis (H&E). The anchorage of the epidermis to dermis is represented by an uneven layer characterized by numerous finger-like protrusions called dermal papillae. Such mode of layer-to-layer attachment results in an increased interface between these layers (courtesy of Lutz Slomianka, PhD).

At the base of the reticular layer of dermis is a layer of adipose tissue, termed *panniculus adiposus*. Its function is to store energy and provide thermal insulation. The adipose layer is particularly well developed in individuals living in cold climates. In association with a loose connective tissue present below it, panniculus adiposus forms the hypodermis or subcutaneous layer.

2.3.2 Repair, Healing, and Renewal

Dermal repair due to an incision or a laceration is achieved in two stages. First, the damaged collagen fibers at the wound site are removed through the macrophage activity; then, the local proliferation of fibroblasts and subsequent collagen deposition, along with other extracellular matrix components, lead the healing process. Application of sutures reduces the extent of the repair area through maximal closure of the wound, thus minimizing scar formation. A surgical incision is typically made along cleavage lines, parallel to the collagen fibers in an attempt to minimize scarring.

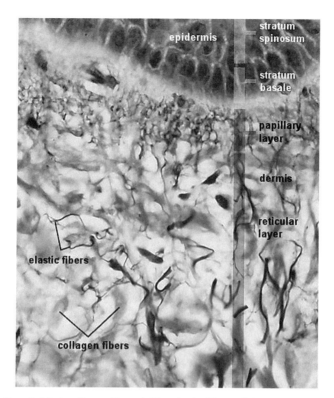

Figure 2.3. Dermis (Verhoeff-van Gieson). The elastin fibers of the papillary layer are thread-like and form an irregular network. In contrast, the elastin fibers of the reticular layer have clear directionality forming regular lines of tension in the skin (courtesy of Lutz Slomianka, PhD).

Repair of the epidermis involves the proliferation of the basal keratinocytes within the undamaged stratum germinativum surrounding the wound. Enhanced mitotic activity within the first 24 hours leads to the formation of a scab. The proliferating basal cells undergo an active process of migration beneath the scab and across the wound surface. It is estimated that the cellular migration rate may be up to 0.5 mm/day, starting within 8–18 hours postinjury [Ross et al. 2003]. Further proliferation and differentiation occur behind the migration front, leading to the restoration of the multilayered epidermis. As a result of cellular keratinization and desquamation, the overlying scab is freed (Figure 2.4).

Diabetes mellitus (DM) represents one of the most critical public health problems worldwide due to its escalating prevalence leading to enormous socioeconomical consequences. It has been reported that diabetes affects 170 million people worldwide, including 20.8 million in the United States, and this number is expected to double by 2030 [Wild et al. 2004]. Factors such as ethnicity, longer life expectancy, epidemiological and nutritional culture, which brings obesity and sedentarism, have been recognized

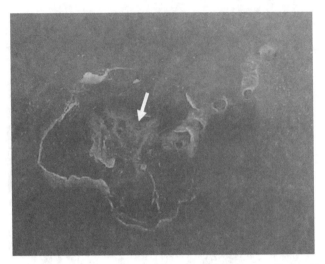

Figure 2.4. Skin repair. The fibroblasts synthesized and deposited collagen fibers at the site of the injury. The maturation of collagen fibers over time, under the scab, leads to wound contraction. This scabless image illustrates the orientation of the collagen fibers at the site of repair (white arrow).

as culprits [Yusuf et al. 2001; Aschner 2002]. Diabetic foot ulcers (DFUs) are one of the chronic consequences of DM representing the most important cause of nontraumatic amputation of the inferior limbs [Silva et al. 2007]. DFU are estimated to occur in 15% of the diabetic population and are responsible for 84% of the diabetes-related lower leg amputations [Brem & Tomic-Canic 2007; Silva et al. 2007]. DFU is a consequence of three fundamental complications associated with diabetes: peripheral neuropathy, vascular insufficiency, and infections [Anandi et al. 2004]. Predisposition toward planar pressure points, foot deformities, and minor trauma further complicate DFU.

Peripheral neuropathy represents the most common complication affecting the lower extremities of patients diagnosed with DM, occurring in more than half of the diabetic population [Armstrong & Lavery 1998]. The lack of protective sensation escalated by the presence of foot deformities exposes patients to sudden or repetitive stress leading to ulcer formation, infection, and, ultimately, amputation [Armstrong & Lavery 1998]. The pathology of the diabetic foot is characterized by the denervation of the dermal structures impairing normal sweating. This is followed by the development of dry, cracked skin which is predisposed to infection. Vascular insufficiency, particularly peripheral arterial occlusive disease (PAOD), is another factor responsible for the development of DFU. PAOD is four times more prevalent in diabetic patients and leads to lower extremity ischemia [Armstrong & Lavery 1998]. Smoking, hypertension, and hyperlipidemia commonly contribute to the increased prevalence of PAOD [Brem & Tomic-Canic 2007; Silva et al. 2007]. Infection complicates the pathological picture of the diabetic foot and plays the main role in the development of moist gangrene. Bacteria from the *Pseudomonas*, *Enterococcus*, and *Proteus* species have been

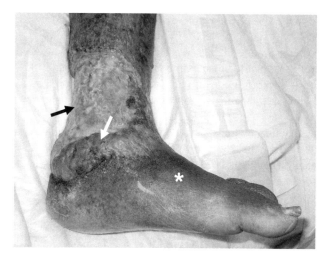

Figure 2.5. Diabetic ulcer. Diabetic patient with a left leg ulcer characterized by significant skin loss of the lower leg (black arrow), hypertrophic granulation tissue at the bottom of the ulcer (white arrow), and evident signs of the peripheral arterial occlusive disease leading to necrosis of the foot (white asterisk) (courtesy of Department of Hospital Medicine, Clinical Cases and Images blogspot, Cleveland Clinic, Cleveland, Ohio).

identified in DFU infections, being responsible for continuous and extensive tissue destruction [Anandi et al. 2004].

The wound healing process in DFU is impaired due to the presence of the DM-induced pathophysiology. The normal wound healing response of the skin has three identifiable stages: inflammation, collagen synthesis and deposition (fibroplasia), and remodeling, ultimately leading to scar formation. In contrast, the chronic skin ulcers, such as DFU, are characterized by prolonged inflammation, defective remodeling of the extracellular matrix, and failure to re-epithelialize. The causes of such outcome cycle back to the main factors leading to DFU, specifically, lack of angiogenesis as a result of vascular occlusion, impaired deposition of the extracellular matrix due to fibroblast phenotypic change as a consequence of peripheral neuropathy, and absence of contraction and epithelialization, again resulting from vascular occlusion and peripheral neuropathy (Figure 2.5) [Fu 2005; Stojadinovic et al. 2005; Brem & Tomic-Canic 2007].

Bioengineered skin products such as cultured epidermal grafts (Epicel®), dermal substitutes (AlloDerm® and Dermagraft®), dermal and synthetic epidermal substitutes (Integra®) have revolutionized the field of skin wound treatment. However, the complicated pathology presented by DFU allows for only temporary relief through the use of such products, aside from the expensive cost of these skin alternatives [Fu 2005; Silva et al. 2007].

Massive destruction of all the epithelial structures of the skin precludes re-epithelialization. These wounds could be healed only by grafting epidermis to cover the wound area. In the absence of a graft, in the best of circumstances, the wound would re-epithelialize slowly and imperfectly by ingrowth of cells from the margins of the wound. Composite synthetic or biological dressings are often used to speed wound repair and improve the quality of healing in chronic or burn wounds [Clark et al. 2007]. Although effective, these solutions do not offer a permanent treatment. Eventually, a split-thickness autograft or allograft may be required to achieve complete healing. However, tissue harvest and transplantation have associated risks such as donor site morbidity or rejection of the transplant. Thus, substitution of split-thickness autografts with autologous epidermal sheets grown from small punch biopsies have been used to facilitate the repair of both epidermal and partial thickness wounds [Clark et al. 2007]. Recently, cultured epidermal sheet allografts have been developed to overcome the necessity to biopsy each patient and delay the treatment by 2–3 weeks required to allow the development of the epidermal harvest into an autograft product. The cell harvest source of most current allografts is neonatal foreskin keratinocytes which are more responsive to mitogens than adult (cadaveric) cells. These allografts have been found to promote accelerated healing and become a reliable treatment in a variety of acute and chronic skin ulcers without evidence of immunological rejection. During a period of a few weeks, the neonatal keratinocytes are replaced by autologous cells [Clark et al. 2007].

Autogenic or allogenic epidermal replacement fails to produce a satisfactory response in full-thickness wounds due to lack of mechanical strength and susceptibility to contracture [Clark et al. 2007]. Further, current skin substitutes lack several skin cells including mast cells, Langerhans cells, and adnexal structures. Although some of these structures may not be necessary for patient survival, they are important for the restoration of normal skin functions such as sensation and sweating [Andreadis 2007]. Consequently, while significant progress has been achieved in developing reliable and long-lasting treatments to skin repair, providing a regenerative solution, especially for a complex pathology such as that presented by DFU, remains a challenge.

2.4   MUSCLE TISSUE

2.4.1   Background

Approximately 40% of the body is composed of skeletal muscle, and an estimated 10% is represented by the smooth and cardiac muscles. The dominant function of the muscle tissue is to generate motile forces through contraction. Such function is achieved through aggregates of specialized, elongated cells arranged in parallel arrays. The contractile forces develop from the internal contractile proteins (actin and myosin). There are two types of contractile muscle filaments (myofilaments): thin filaments (6–8 nm in diameter, 1 μm long) composed primarily of the protein actin, and thick filaments (~1.5 nm in diameter, 1.5 μm long) composed of the protein myosin II [Young 2000; Ross et al. 2003].

The cytoarchitectural arrangement of the thin and thick filaments as it appears in histological preparations allows for visual distinction and classification of the muscle tissue into three categories (Figure 2.6):

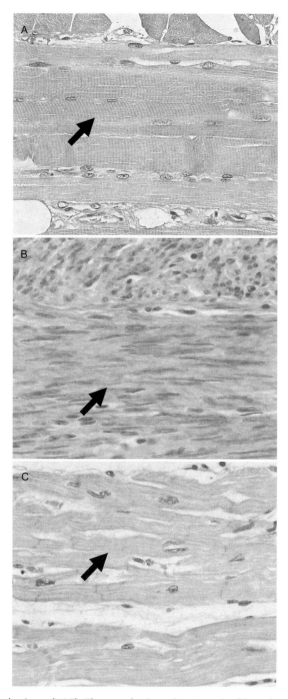

Figure 2.6. Muscle tissue (H&E). The muscle tissue is categorized based on its cytoarchitectural arrangement of thin and thick filaments into (A) skeletal muscle (black arrow), (B) smooth muscle (black arrow), and (C) cardiac muscle (black arrow) (histological images courtesy of Lutz Slomianka, PhD).

- skeletal muscle,
- smooth muscle,
- cardiac muscle.

2.4.2   Skeletal Muscle

2.4.2.1   Structure and Functionality. The skeletal muscle is responsible for the movement of the axial and appendicular skeleton and for the maintenance of body position and posture. Further, it is responsible for the movement of organs such as the globe of the eye and the tongue. Since it is capable of voluntary control, the skeletal muscle is often referred to as the *voluntary muscle*. Due to the striated appearance of its cytoarchitectural arrangement in histological preparations, the skeletal muscle is also termed *striated muscle* [Young 2000].

The skeletal muscle is composed of elongated, multinucleated contractile cells bound together by lamina. The individual muscle fibers vary considerably in diameter from 10 to 100 μm and, in some cases, may extend throughout the entire length of a muscle reaching up to 35 cm in length [Young 2000]. The portion of the muscle fiber that lies between two successive Z disks is called a sarcomere. It represents the basic unit of a striated muscle (Figure 2.7). Scanning electron microscopy (SEM) represents a powerful imaging tool used to examine biological structures beyond the capabilities of traditional optical imaging. In this case, SEM was used to illustrate the structure of the sarcomere. During contraction, the I band and H zone shorten, the distance between the two Z lines of the sarcomere shortens, while the A band maintains its original length.

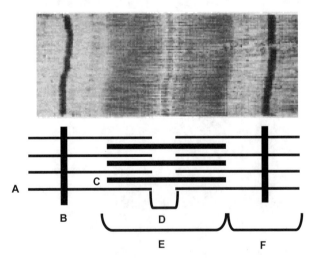

Figure 2.7. The sarcomere of the skeletal muscle (SEM). The sarcomere represents the basic contractile unit of the skeletal muscle featuring the following components: (A) thin filaments (actin), (B) Z line, (C) thick filaments (myosin), (D) H band, (E) A band, and (F) I band.

A

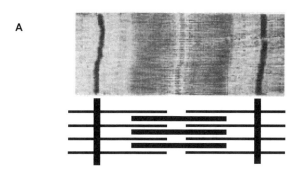

B

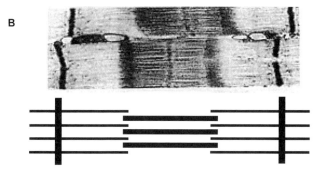

Figure 2.8. Muscle activity (SEM). During stretching, the width of the I band and H zone increases, along with the distance between the Z lines, while the A band maintains its original length: (A) sarcomere in relaxed state and (B) sarcomere during muscle stretching.

Conversely, as the muscle stretches, the width of the I band and H zone increases, along with the distance between the Z lines, while the A band maintains its original length (Figure 2.8).

The skeletal muscle contraction is controlled by large motor nerves. Individual nerve fibers branching within the muscle supply a group of muscle fibers, collectively described as a *motor unit* (Figure 2.9). A silver preparation was used to visualize the neuronal structures (axons and motor plate, black color) within Figure 2.9. The excitation of a motor nerve results in the simultaneous contraction of all the muscle fibers corresponding to the motor unit. The presence of a nerve supply is fundamental to the vitality of the skeletal muscle fibers which will atrophy if the nerve supply is damaged. The skeletal muscle only contracts when an impulse is received from a motor neuron. During stimulation of the muscle cell, the motor neuron releases acetylcholine, a neurotransmitter, which travels across the neuromuscular junction (the synapse between the terminal bouton of the neuron and the muscle cell) until it reaches the sarcoplasmic reticulum. The action potential from the motor neuron changes the permeability of the

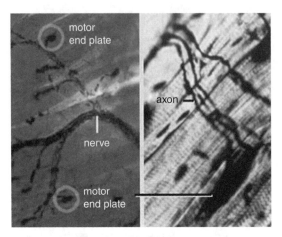

Figure 2.9. The motor unit (Silver stain preparation). The motor unit consists of the nerve branches and the muscle fibers they innervate. The contact area between the muscle fibers and the motor nerve is known as the motor end plate (courtesy of Lutz Slomianka, PhD).

sarcoplasmic reticulum, allowing the flow of calcium ions into the sarcomere, thus leading to muscle contraction.

2.4.2.2 Repair, Healing, and Renewal.
In skeletal muscle development, myoblasts fuse to form multinucleated myofibers. The regenerative capacity of the skeletal muscle is supported by skeletal muscle fiber interposed *satellite cells* which function as site-specific stem cells. Following injury, the satellite cells proliferate, giving rise to new myoblasts. This approach to skeletal muscle regeneration is limited [Ross et al. 2003].

Myosin, actin, tropomyosin, and troponin form on average 75% of the proteins in the muscle fibers. Their fundamental function in attaching and organizing the filaments within the sarcomere, and connecting the sarcomeres to the plasma membrane and the extracellular matrix render them indispensable to maintaining the overall function of the muscle. Mutations in the genes encoding these proteins may produce defective proteins which could have a devastating effect on one's health. One of the most severe and, for that reason, most documented mutations occurs in the dystrophin gene, leading to Duchenne muscular dystrophy, a genetic disease characterized by progressive degeneration of skeletal muscle fibers (Figure 2.10). The excessive demand placed on the satellite cells to replace the degenerated fibers leads to exhausting the stem cell pool. Even though the body attempts to recruit additional myogenic cells from the bone marrow, the rate of degeneration surpasses the rate of regeneration, thus resulting in loss of muscle function [Ross et al. 2003]. Patients inflicted by this disease succumb to their death as a result of compromised respiratory function.

Presently, there is no clinical treatment to stop or reverse muscular dystrophy. However, multiple gene therapy strategies have reached preclinical evaluation stage. The scientists anticipate that optimization of vascular delivery routes using surrogate,

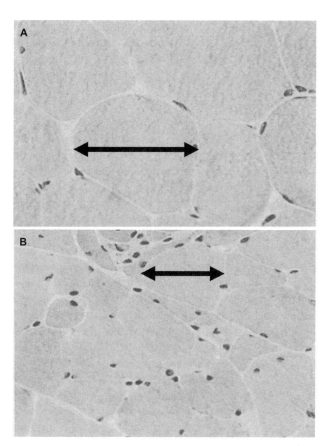

Figure 2.10. Duchenne muscular dystrophy (H&E). Cross section of skeletal muscle fibers at identical magnification: (A) normal muscle tissue and (B) degenerated muscle tissue due to muscular dystrophy. A significant decrease in muscle fiber diameter is evident in (B) compared to (A) (black double head arrow) (histological images courtesy of James M. Anderson, MD, PhD).

replacement, or booster genes could lead to clinically meaningful results [Rodino-Klapac et al. 2007].

2.4.3 Smooth Muscle

2.4.3.1 Structure and Functionality. This muscle type forms the muscular component of the viscera and the vascular system, the arrector pili muscles of the skin, and the intrinsic muscles of the eye. In the case of the smooth muscle, the cytoarchitectural arrangement of contractile proteins in histological preparations does not give the appearance of cross striations (*nonstriated muscle*). This is because the myofilaments do not achieve the same degree of order in their arrangement as in the skeletal muscle cytoarchitecture [Ross et al. 2003] (Figure 2.6B).

The smooth muscle is composed of fibers that range in length from 20 μm in the walls of the small blood vessels to approximately 200 μm in the wall of the intestine; they may be as long as 500 μm in the wall of the uterus during pregnancy [Ross et al. 2003]. In contrast to the skeletal muscle, specialized for voluntary, short duration contractions, smooth muscle is specialized for rhythmic or wave-like contractions producing diffuse movements that result in the response of an entire muscle mass rather than an individual motor unit. Contractility occurs independent of the neurological innervation. Tension generated by contraction is transmitted through the anchoring densities in the cytoplasm and the cell membrane to the surrounding external lamina, thus allowing a mass of smooth muscle cells to function as one unit [Young 2000]. Superimposed on the inherent contractility are the influences of the autonomic nervous system, hormones, and local metabolites that modulate contractility to accommodate changes in functional demands. Since the smooth muscle is under inherent autonomic and hormonal control, it is described as an *involuntary muscle*.

2.4.3.2 Repair, Healing, and Renewal. Smooth muscle cells are capable of dividing to maintain or increase their number. They respond to injury by undergoing mitosis during uterine smooth muscle cell proliferation as part of the normal menstrual cycle and pregnancy, as well as to replace damaged or senile muscular cells of blood vessels [Ross et al. 2003]. As the predominant cellular element of the vascular media, smooth muscle cells play a fundamental role in the normal functioning of blood vessels.

Coronary artery disease (CAD) is the leading cause of morbidity and mortality in the United States. In 2003, American Hear Association statistics reported over 650,000 deaths attributed to CAD. Further, CAD represents the largest health-care expenditure for a single disease in the United States [Weintraub 2007]. One of the leading advances for the treatment of CAD emerged with the advent of percutaneous transluminal coronary angioplasty (PTCA) allowing reperfusion of the affected (blocked) coronary segment. However, it has been reported that 30–50% of the patient population following prescribed simple balloon angioplasty or stent placement experiences *restenosis* [Cotran et al. 1999; Weintraub 2007].

The mechanism of restenosis is incompletely understood. It has been established that trauma induced during angioplasty consisting of thrombosis, inflammation, and extracellular matrix deposition within the lumen of the vessel, and extensive smooth muscle cell proliferation contributes to the lumen narrowing within 6 months postintervention [Nikol et al. 1996; Weintraub 2007] (Figure 2.11). In particular, neointimal hyperplasia leading to luminal narrowing is the result of smooth muscle cell migration at the site of injury and their production and deposition of extracellular matrix. The smooth muscle cell proliferation is triggered by a complex sequence of events initiated immediately posttreatment. Following angioplasty, the thrombogenic components of the fractured plaque are exposed to the circulating blood leading to platelet adhesion and activation, then thrombosis. Loss of endothelium, the antithrombogenic aspect of the vessel in contact with the blood, encourages further platelet adhesion and thrombosis. The mitogens released by the activated platelets (thromboxane, serotonin, platelet-derived growth factor, etc.) then initiate the proliferative phenomenon experienced by the smooth muscle cells, altering their phenotype from contractile to synthetic

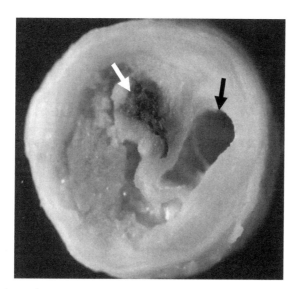

Figure 2.11. Atherosclerosis is a disease affecting the arterial blood vessels. It results in clogging, hardening, and narrowing of these blood conduits due to the deposition of lipoproteins leading to the formation of an atheromatous plaque. The plaque is composed of the atheroma (yellow, flaky material inflammatory in nature), cholesterol crystals near the lumen of the vessel, and, in the case of old lesions, calcification at the outer perimeter of the vessel. The featured plaque also presents a thrombus within the atheroma (white arrow), and shows a reduction in lumen diameter of 75% (black arrow) (image courtesy of James M. Anderson, MD, PhD).

(production of extracellular matrix). In return, the smooth muscle cells secrete promigratory proteins which stimulate further migration of the medial smooth muscle cells to the location of platelet adhesion/thrombosis [Miano et al. 1993; Pakala et al. 1997; Weintraub 2007].

The Taxus® stent technology using a naturally occurring mitotic inhibitor (paclitaxel) is presently attempting to minimize/eliminate the restenotic event related to smooth muscle cells by taking advantage of pharmacological advances through which the smooth muscle cell proliferation could be selectively controlled by microtubule stabilization, which results in inhibition of G0/G1 and G2/M phases of cellular division [Chen et al. 2007; Daemen & Serruys 2007; Pires et al. 2007]. However, permanent physical irritation, thrombogenicity, and obstruction of normal remodeling have been identified as limitations of the present technology (Figure 2.12). In an attempt to eliminate the aforementioned shortcomings, the realm of stent design has been opened recently to the use of biodegradable polymers associated with encapsulated controlled dose release therapeutic drugs. The success of such approach requires thorough evaluation in future randomized clinical trials.

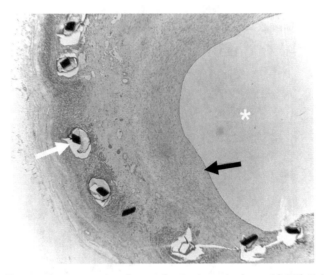

<u>Figure 2.12.</u> Restenosis, cross-sectional area of a stent repaired vessel (H&E). Following place-ment of the stent, a restenotic event led to the aggressive full encapsulation of the device, the white arrow indicates the location of the stent struts, the black arrow indicates the new lumen line in contact with blood, significantly advanced compared to the location of the stent, and the asterisk indicated the blood circulating area (histological image courtesy of James M. Anderson, MD, PhD).

2.4.4　Cardiac Muscle

2.4.4.1　Structure and Functionality. The cardiac muscle is found in the wall of the heart and at the base of the large veins that empty into the heart. The cardiac muscle has many functional characteristics, *intermediate* between those of the skeletal and smooth muscle, and ensures the continuous, rhythmic contractility of the heart. Although the cytoarchitecture of the cardiac muscle (*striated muscle*) is similar to that of the skeletal muscle, it is often difficult to visualize the striated aspect in histological prepa-rations [Ross et al. 2003] (Figure 2.13).

Unlike skeletal and smooth muscle fibers which are characterized by multinucle-ated single cells, cardiac muscle fibers consist of numerous cells arranged end to end. In addition, in a fiber, some cardiac muscle cells may join with adjacent cells through *intercalated disks*, thus creating a branched fiber. The branching fiber arrangement typical to the cardiac muscle is invoked in explaining the difficulty in visualizing the muscle striations (Figure 2.13). The section featured in Figure 2.13 takes advantage of the Alizarin Blue to illustrate the cardiac muscle striations.

In the case of the cardiac muscle, the sarcoplasmic reticulum allows for a slow leak of calcium ions into the cytoplasm after recovery from the preceding cardiac contraction. This specific aspect of the cardiac muscle is responsible for a succession of automatic contractions independent of external stimuli. The rate of this inherent rhythm is then modulated by autonomic and hormonal cues [Young 2000; Ross et al.

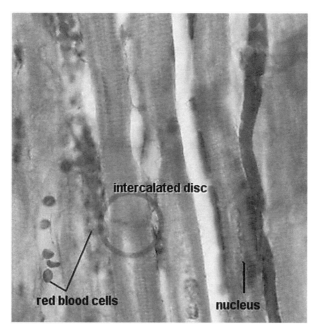

Figure 2.13. Cardiac muscle (Alizarin Blue, Whipf's Polychrome, Iron Haematoxylin, H&E). The cytoarchitecture of the cardiac muscle is dissimilar from that of the skeletal muscle. Through intercalated disks the cardiac cells form branched fibers (courtesy of Lutz Slomianka, PhD).

2003]. Further, the intercellular junctions termed as intercalated disks provide points of anchorage for the myofibrils and permit rapid spread of the contractile stimuli from one cell to another (Figure 2.13). Therefore, adjacent fibers are caused to contract almost simultaneously, thereby acting as a functional muscle mass. Finally, a system of highly modified cardiac muscle cells constitutes the *pacemaker region* of the heart and ramifies throughout the organ as the Purkinje system, thus coordinating the contraction of the cardiac muscle as a whole during each cardiac cycle [Ross et al. 2003].

2.4.4.2 Repair, Healing, and Renewal. Mature cardiac muscle cells do not divide under normal conditions. Traditionally, the death of cardiac muscle cells due to a localized injury leads to the formation of fibrous connective tissue devoid of any cardiac function. Such pattern of injury and repair is typical to *myocardial infarction* (Figures 2.14 and 2.15). However, a study of transplanted hearts removed from recipients has identified a low rate (0.1% of the population) of cellular mitosis [Beltrami et al. 1997].

The concept of cell transplantation for cardiac repair has been introduced nearly a decade ago. In vitro engineered cardiac tissue has been developed, with mixed results, by a number of research groups using specialized bioreactor technology to create a functional cardiac tissue scaffolds [Christman & Lee 2006; Engelmann & Franz 2006; Kubo et al. 2007]. Translation of this research to clinical level is the focus of intense ongoing collaborative effort between clinicians, cell biologists, and material scientists.

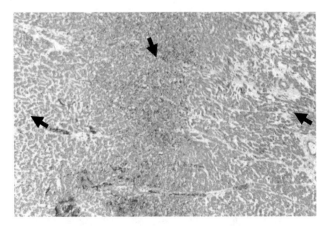

Figure 2.14. Acute myocardial infarction (H&E). This section illustrates the edge of the infarct with normal tissue to the left and the infracted tissue to the right (black arrows). The central strip of the section shows the presence of inflammatory cells in the proximity of the infarcted tissue (courtesy of IPLAB.net).

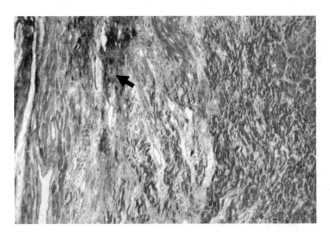

Figure 2.15. Healed myocardial infarction (Masson's TriChrome Blue). To the left of the section the presence of the dense connective tissue (scar) is evident (black arrow). To the right of the image, normal myocardial tissue is present (courtesy of IPLAB.net).

2.5 CONNECTIVE TISSUE

2.5.1 Background

The connective tissue is widespread in the body. The principal roles of the connective tissue are to bind or strengthen organs or other tissues. It also functions inside the body to divide and compartmentalize other tissue structures bounded by basal laminae of the

TABLE 2.3. Classification of Connective Tissues

Classification	Examples
Embryonic connective tissue	• Mesenchyme
	• Mucous connective tissue
Connective tissue proper	• Dense connective tissue (regular or irregular)
	• Loose connective tissue
Specialized connective tissue	• Adipose tissue
	• Blood
	• Bone
	• Cartilage
	• Hematopoietic tissue
	• Lymphatic tissue

various epithelia and by the basal of external lamina of muscle cells and nerve supporting cells. Connective tissues are composed of *cells* and extracellular matrix that includes *fibers* and *ground substance*.

The functions of various connective tissues are reflected by the type of cells and fibers present within the tissue and the composition of the ground substance of the extracellular matrix. The cells are not close together as in the epithelium, but they are spread out at intervals along the connective tissue fibers and embedded into the ground substance. The ground substance consists of a variety of glycosaminoglycans (GAGs), such as chondroitin sulfate and hyaluronic acid, in combination with other proteins. The classification of connective tissue is based on the composition and organization of its cellular and extracellular components and on its function. Consequently, the connective tissue has been classified as follows: embryonic connective tissue, connective tissue proper, and specialized connective tissue. Each category is further subdivided into subsequent subcategories as illustrated in Table 2.3 [Ross et al. 2003].

2.5.2 Embryonic Connective Tissue

The embryonic connective tissue is classified into mesenchyme and mucous connective tissue.

Mesenchyme, a primitive connective tissue, is established in the embryo. Mesenchymal cells migrate from the mesoderm into the spaces between the germ layers to form a loosely arranged three-dimensional cellular network. Mesenchymal maturation gives rise to the various connective tissues in the body such as tissue proper, cartilage, bone, blood, smooth muscle, and endothelium. The in situ stimuli specific to each tissue proper are responsible for the specific differentiation path adopted by the mesenchyme. Some mesenchymal cells persist through adulthood and serve as a cellular repository from which highly specialized cells can develop [Ross et al. 2003]. One such example is represented by the differentiation of a mesenchymal cell into a hemocytoblast, the stem cell from which blood cells are derived.

Mucous connective tissue is present in the umbilical cord (Figure 2.16). It consists of a specialized extracellular matrix whose ground substance is frequently referred to as

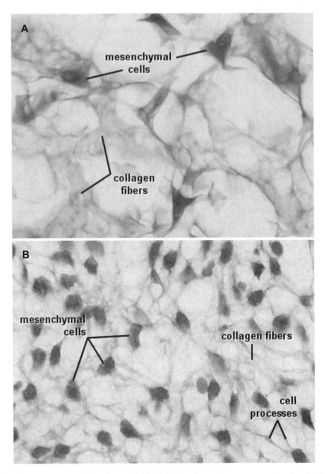

<u>Figure 2.16.</u> Mucous connective tissue (H&E). This tissue is composed of gelatin-like ground substance, mesenchymal cells, and collagen fibers. The collagen fibers form a very open structural network consistent with the limited physical stress experienced by this tissue. The cells and fibers are embedded within the gelatin-like ground substance which, during histological tissue preparation, is lost: (A) umbilical cord and (B) fetal kidney (courtesy of Lutz Slomianka, PhD).

Wharton's jelly due to its gelatin-like appearance. Most of the cells in the mucous connective tissue are active fibroblasts. These cells are responsible for the generation of the fibers and ground substance that lie between the cells [Ross et al. 2003]. In the mature adult, the mucous connective tissue is present solely in the vitreous body of the eye.

2.5.3 Connective Tissue Proper

The components of connective tissue proper are cells, fibers, and ground substance. The cells and fibers are embedded into the soft ground substance.

2.5.3.1 Cells of the Connective Tissue Proper. The cells associated with the connective tissue proper can be categorized according to their basic function into [Young 2000]

- cells responsible for the synthesis and maintenance of the extracellular matrix, such as the resident fibroblast or resident myofibroblast population;
- cells responsible for the storage and metabolism of fat, known as resident adipocytes;
- cells specialized in defense against an insult (mechanical injury, immune response, etc.), represented by the resident macrophages and mast cells, as well as the transient cell populations arriving from the blood in response to such an insult.

The resident fibroblast population is responsible for the synthesis of collagen, elastic and reticular fibers, and complex carbohydrates of the ground substance. The cellular machinery of each fibroblast is capable of synthesizing all the necessary protein components of the extracellular matrix. Fibroblasts are embedded among the collagen fibers of the connective tissue proper and play a fundamental role in the process of repair, renewal, and differentiation (Figure 2.17).

Adipocytes represent the basic cell unit of the adipose tissue. They are responsible for the storage of lipids. The mature adipocyte is characterized by a single, large lipid inclusion surrounded by a thin rim of cytoplasm (Figure 2.18A). The lipid inclusion compresses the nucleus against the cytoplasmic aspect of the cellular membrane (Figure 2.18B). The lipid inclusion is lost during histological preparation of respective tissue sections. The storage and metabolism of fat is under sympathetic and hormonal control. Such control is further enhanced through the adipocytes' endocrine function which relays information concerning the fat reserves of the body to the brain.

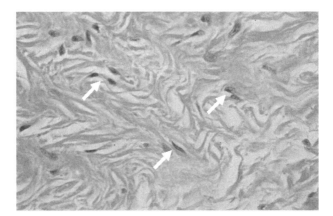

Figure 2.17. Fibroblasts (H&E). The fibroblasts are embedded between the collagen fibers and the ground substance of the tissue (white arrows) (courtesy of Ed Friedlander, MD).

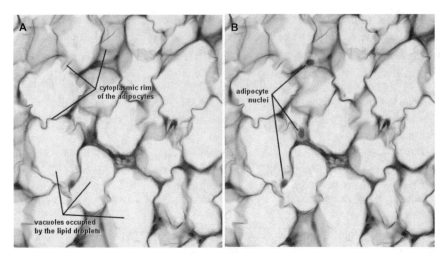

Figure 2.18. Adipocytes (H&E). The large lipid vacuoles occupy the central portion of the mature adipocyte compressing the nucleus against the cell membrane: (A) and (B) represent an identical image of the human adipose tissue introduced twice for ease of labeling (courtesy of Lutz Slomianka, PhD).

Macrophages are phagocytic cells critical in mounting a defense response against an insult as well as being instrumental in debris removal (Figure 2.19). Their precursor, the monocyte, a circulating blood cell, migrates from blood into the connective tissue where it differentiates into a macrophage. The role of macrophages in the immune response reactions has been thoroughly documented. The Major Histocompatibility Complex II proteins present on the surface of macrophages allow them to interact with helper CD4$^+$ T lymphocytes. In the process of foreign body phagocytosis, the macrophage displays a short polypeptide (antigen) specific to the engulfed foreign body, thus triggering the activation of the CD4$^+$ T lymphocyte which initiates the immune response [Abbas et al. 2000; Ross et al. 2003].

2.5.3.2 Connective Tissue Proper Fibers.
The connective tissue fibers are classified into collagen fibers (including reticular fibers) and elastic fibers. Collagen fibers represent the most abundant type of connective tissue fiber. Further, collagen accounts for 30% of the total human body protein. One of collagen's most notable functions is its tensile strength adapted to a plethora of support requirements delineated by the different types of specialized connective tissues (Table 2.3).

Tropocollagen, the basic collagen unit, is a rod-like structure of about 300 nm long and 1.5 nm in diameter. It comprises three left-handed polypeptide strands giving rise to a right-handed coil, known as a triple helix. Individual tropocollagen units undergo a process of self-assembly into large arrays of various architectures, densities, amino-acid composition, and strengths depending on the in situ support requirements. As with any protein, hydrogen bonds stabilize the quaternary structure of collagen. Covalent cross-linking between different tropocollagen helices represents one of the factors

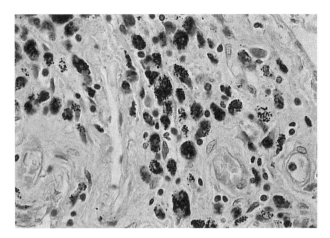

Figure 2.19. Macrophages (H&E). The cells, also known as mononuclear phagocytes, play a significant role in the inflammatory and wound healing response as well as the immune response of the body. In this preparation, the macrophages are identified as a result of their ingestion of carbon particles (foreign bodies) present in a smoker's lung (courtesy of Ed Friedlander, MD).

which accounts for the variety in collagen types. Collagen fibrils represent tropocollagen units packed into an organized overlapping bundle. At the next hierarchical level, bundles of fibrils form organized collagen fibers.

The polypeptide that gives rise to the triple helix is composed of a repeating array of amino acids often following the pattern glycine-X-proline or glycine-X-hydroxyproline, where X could be any of the naturally occurring amino acids. The high content of proline and hydroxyproline rings, with their geometrically constrained carboxyl and secondary amino groups, accounts for the tendency of the individual polypeptide strands to form left-handed helices spontaneously, without any intrachain hydrogen bonding. Glycine, as the smallest amino acid, plays a unique role in collagen. Glycine is required at every third position because the assembly of the triple helix places this residue at the interior of the helix, where there is no space for a larger side group than glycine's single hydrogen atom. To further accommodate such constraint, the rings of the proline and hydroxyproline point outward [Cantor & Schimmel 1980].

Collagen is secreted by fibroblasts into the extracellular matrix in the form of tropocollagen which, in the extracellular matrix, polymerizes into collagen. To date, at least 19 different types of collagen have been identified. The most notable types of collagen present in the human body are outlined below [Young 2000; Ross et al. 2003] (Figure 2.20):

- Type I collagen is the most abundant type of collagen in the body. It is found in the fibrous supporting tissue, the dermis of the skin, tendons and ligaments, as well as in the hard tissue. It also appears in the scar tissue following soft tissue repair. The tropocollagen units form aggregates of fibrils strengthened by

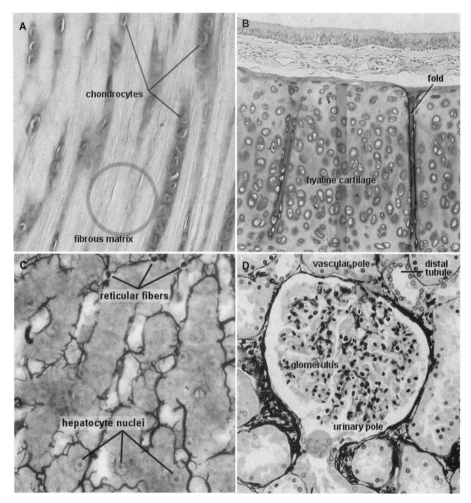

Figure 2.20. Types of collagen: (A) type I collagen predominates in the fibrous cartilage, no imaging distinction apparent between type I and type II collagen here (intervertebral disk, H&E), (B) type II collagen is present in the hyaline cartilage (trachea, H&E), (C) type III collagen (reticulin) is present in the high cell density environments (liver, reticulin stain—black), (D) type IV collagen is part of filtration barriers (kidney glomerulus, silver methenamine stain—black fibers around the glomerulus) (courtesy of Lutz Slomianka, PhD).

numerous intermolecular bonds. Its architecture and organization differs from tissue to tissue based on individual support requirements, thus giving rise to loose and dense connective tissue.

• Type II collagen is exclusively present in the hyaline cartilage and is composed of fine fibrils dispersed into the ground substance.

- Type III collagen is also known as reticulin. The reticulin fibers are predominant in high cell density environments (liver, intestine, uterus, lymphoid organs) where they form a delicate branched support network. The reticulin fibers also fulfill a transitory function within the wound healing response continuum. They represent the collage of the granulation tissue produced by fibroblasts as a response to injury. Over time, the reticulin fibers mature and organize into type I collagen fibers, the hallmark protein of the scar tissue.
- Type IV collagen forms a mesh-like structure rather than fibrils, and it is an important constituent of the basement membranes. It is localized in the basal lamina and the lens of the eye. It also serves as a filtration barrier within the glomeruli of the nephron.
- Type VII collagen is present in anchoring fibrils that link to the basement membrane, such as the epidermal junctions.

The remainder of the collagen types is present in various specialized situations.

Elastic fibers form a three-dimensional network, and they are typically thinner than the collagen fibers. In contrast to collagen, the elastic fibers have two structural components: a central core of elastin and surrounding fibrillin microfibrils [Ross et al. 2003]. Elastin is primarily composed of the amino acids glycine, valine, alanine, and proline. Through the linking of many tropoelastin protein molecules, a large lamellar, durable cross-linked array is formed, such as that found in blood vessels (Figure 2.21). Desmosine and isodesmosine are two large amino acids directly responsible for the random coiling of elastin. The achievement of random coiling is responsible for the elastic properties of these fibers which, following stretching, return to their original relaxed state. The role of fibrillin is to organize the newly deposited elastin into fibers during elastogenesis [Ross et al. 2003]. Then, fibrillin becomes incorporated around and within the elastic fibers [Young 2000].

2.5.3.3 Ground Substance.
The ground substance occupies the space between the cells and the fibers of the connective tissue. It is a semifluid gel which facilitates the fluid transport throughout the connective tissue. The ground substance consists mainly of proteoglycans, large macromolecules composed of a core protein to which GAGs are covalently bound [Ross et al. 2003]. GAGs are long-chain unbranched polysaccharides (sugars). GAGs are intensely negatively charged due to the presence of hydroxyl, carboxyl, and sulfate side groups on the disaccharide units. A consequence of the localized negative charge is the inherent hydrophilicity of the ground substance leading to its native semifluid (gel) aspect [Ross et al. 2003]. One notable exception is hyaluronic acid. This GAG does not display sulfate side groups; thus, it is more rigid than most of the GAGs due to its reduced propensity of interacting with water. In addition, it is characterized by a significantly longer polysaccharide chain on the order of thousands of sugars, compared to several hundreds or less sugar units present in most GAGs. In particular, in the ground substance of the cartilage, proteoglycans indirectly bind to hyaluronic acid forming macromolecular arrays accounting for the ability of cartilage to resist compression without inhibiting flexibility [Ross et al. 2003].

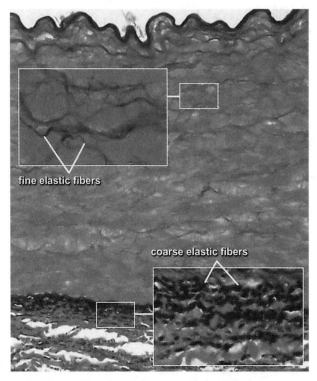

Figure 2.21. Elastic fibers (Verhoeff's&Eosin). The recoil (elastic) property of these fibers is captured by their undulating aspect in large artery sections, where they form the elastic lamina responsible for the dynamic dilation of these vessels (courtesy of Lutz Slomianka, PhD).

2.5.4   Specialized Connective Tissues

Among the specialized connective tissues outlined in Table 2.3, particular attention is given to cartilage.

2.5.4.1   Structure and Function. The cartilage is a specialized dense connective tissue composed of an extracellular matrix populated with cells known as chondrocytes. The extracellular matrix is produced and maintained by resident chondrocytes, and upon reaching maturity, it becomes avascular. The viability of the tissue is ensured through the large ratio of GAGs to extracellular matrix type II collagen which permits adequate diffusion between the chondrocytes and the blood vessels of the surrounding connective tissue [Ross et al. 2003]. A natural design parameter of the cartilage affected by its avascular nature is the thickness to which the matrix could develop in order not to compromise the viability of the diffusion-dependent innermost chondrocytes. The supporting matrix of the cartilage is superiorly adapted to the demanding dynamically active mechanical environment within which it performs. The distinct structure (architecture and composition) of cartilage in response to its various functions has led to the following classification:

- hyaline cartilage,
- fibrocartilage,
- elastic cartilage.

The hyaline cartilage is the most common type of cartilage found in many articular surfaces, nasal septum, larynx and tracheal rings, and the sternal ends of the ribs. It is characterized by a matrix composed of type II collagen, proteoglycans, and hyaluronic acid. The open lattice aspect of the hyaline cartilage ensures a regulated distribution of chondrocytes which reside within the matrix voids. The role of the hyaline cartilage is to provide a low friction surface within the joint environment, to participate in lubrication in synovial joints, and to distribute the applied forces to the underlying skeletal tissue [Young 2000; Ross et al. 2003] (Figure 2.22).

The fibrocartilage is most prevalent in the intervertebral disks, some articular joints (menisci of the knee joint), and in the pubic symphysis (midline cartilaginous joint connecting the pubic bones). Further, the fibrocartilage is present in connection with the dense connective tissue of the joint capsules, ligaments, as well as some of the connections between tendons and anchoring bone [Young 2000]. The composition of the fibrous cartilage is represented by type I and II collagen, proteoglycans, and hyaluronic acid. The type I collagen forms layers of dense collagen fibers oriented in the

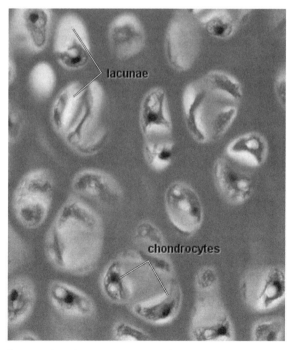

Figure 2.22. Hyaline cartilage (H&E). The matrix of the hyaline cartilage appears amorphous since the ground substance and collagen have similar refractive properties. The aggregation of chondrocytes is embedded into the surrounding matrix (courtesy of Lutz Slomianka, PhD).

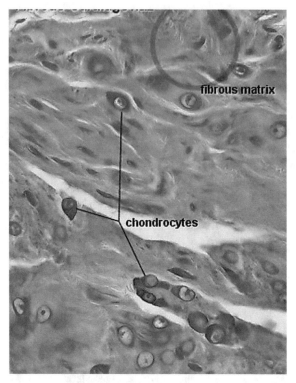

Figure 2.23. Fibrocartilage (H&E). The fibrocartilage is a combination of dense regular connective tissue (type I collagen) and hyaline cartilage (type II collagen). Type I collagen is the predominant collagen component of this connective tissue (courtesy of Lutz Slomianka, PhD).

direction of the functional stress. These layers alternate with hyaline cartilage matrix, providing a composite structure which is pliable as well as resilient (Figure 2.23). The fibrocartilage withstands compression as well as shearing forces, thus acting as a shock absorber.

The elastic cartilage occurs in the external ear and external auditory canal, the epiglottis, parts of the laryngeal cartilage, and the walls of the Eustachian tubes [Young 2000]. Its structure is rich in bundles of elastic fibers interdispersed among the ground substance of the supporting hyaline cartilage. The elastic fibers are predominantly present in the vicinity of the lacunae (Figure 2.24). Unlike typical hyaline cartilage, the elastic cartilage does not calcify with age [Ross et al. 2003].

2.5.4.2 Repair, Healing, and Renewal of Hyaline Cartilage. Degenerative joint diseases represent a leading cause of debilitation, with osteoarthritis alone affecting over 20 million people annually in the United States. The decade of 2002–2010 has been declared by the United Nations and over 160 countries as the Bone and Joint Decade, to promote public awareness, research, and treatment of bone and joint diseases. Over 1 million procedures to treat defects of articular cartilage are performed

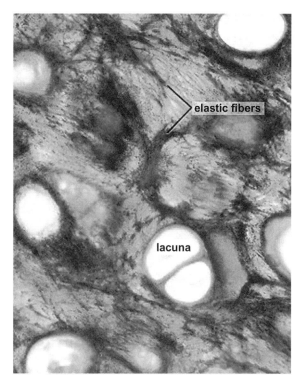

Figure 2.24. Elastic cartilage (Elastin stain, elastic fibers imaged in dark violet). The elastic cartilage can be readily distinguished from the hyaline or fibrocartilage by the presence of elastin in the cartilage matrix (courtesy of Lutz Slomianka, PhD).

annually in the knee alone in the United States [Jackson et al. 2001; Lynn et al. 2004]. Approximately half of these procedures involve treatment of defects that are sufficiently severe to require the replacement or assisted regeneration of damaged hyaline cartilage. Such procedures may provide temporary relief of pain and can delay the need for total knee arthroplasty for several years. However, these procedures result in fibrocartilaginous repair tissue, with its associated mechanical deficiencies leading to breakdown under normal joint loading demands [Shapiro et al. 1993].

The regeneration of articular surfaces presents a unique challenge to scientists due to the limited capacity of this type of cartilage to self-repair [Jackson et al. 2001; Solchaga et al. 2001; Toba et al. 2002; Cancedda et al. 2003]. Current available therapies are primarily directed toward small focal cartilage defects and do not provide lasting or permanent solutions to the severely damaged cartilage. One proposed direction to the replacement or regeneration of articular cartilage involves the use of multiphase compartmental scaffolds which incorporate separate osseous (bone) and cartilaginous components. Such approach aims to address the separate requirements for regeneration of bone (the anchoring base for the scaffold) and the articular cartilage (the functional component) which must simultaneously be met for the development of a healthy articular cartilage [Hunziker et al. 2001; Gao et al. 2002; Schaefer et al. 2002; Sherwood et al. 2002; Hung et al. 2003; Hunziker & Driesang 2003].

A critical component to the de novo cartilage development rests with its mechanical conditioning that has been shown to impact chondrocyte distribution within the normal cartilage tissue. In turn, the cellular distribution affects the compliance of the tissue to the regular mechanical stresses it is subjected to. The physiological loading stimulates changes in the metabolic activity of the cells and the surrounding matrix. It is the mechanical conditioning of de novo cartilage development aspect that presently has not been successfully addressed.

2.6 THE FOREIGN BODY RESPONSE

Inflammation, wound healing, foreign body response, and *fibrosis* are recognized as principal phases of the tissue or cellular host response to soft tissue injury and repair [Anderson 2001; Ratner 2001; Kirkpatrick et al. 2002] (Figure 2.25). Inflammation represents the reaction of the vascularized living tissue to the local injury created at the biomaterial implantation site. Then, the wound healing process leads to the formation of the granulation tissue, characterized by angiogenesis and presence of fibroblasts [Anderson 1993] (Figure 2.26). Finally, the foreign body reaction, represented by macrophages and foreign body giant cells (FBGCs), initiates fibrosis and fibrous encapsulation of the biomaterial. The fibrous capsule surrounds the biomaterial isolating the implant from the local tissue environment [Moussy & Reichert 2000; Voskerician 2004]. The size, shape, and chemical and physical properties of the biomaterial are responsible for variations in the intensity and duration of the inflammatory or wound healing process [Quinn et al. 1995; Gerritsen et al. 2000; Chen et al. 2002; Palmisano et al. 2002; Ward et al. 2002; Ban et al. 2003]. The inflammatory and wound healing response can be divided into several fundamental stages: *acute inflammation, chronic inflammation*, formation of *granulation tissue, foreign body reaction*, and *fibrous capsule development* [Anderson 1996] (Figure 2.25).

The acute phase of the inflammatory response occurs immediately after tissue injury, and is of relatively short duration, lasting from minutes to days, depending on the extent of the injury. It is characterized by plasma protein adsorption and the migration of leukocytes from the microcirculation, including polymorphonuclear leukocytes (PMN), monocytes, and lymphocytes. These inflammatory cells actively migrate from the vasculature in response to chemotactic factors present at the implant site. Protein-rich fluid (exudate) accompanies this cellular movement. Further, increased vascular permeability facilitates this movement resulting in accumulation of cells and exudate, and it is the result of several mechanisms including endothelial contraction, cytoskeletal reorganization, leukocyte-mediated endothelial cell injury, and leakage from regenerated capillaries [Anderson 1996].

The predominant cell type within the exudate during the acute phase of inflammation is the PMN, also called neutrophil. Its major role is to attack and phagocytose bacteria, tissue debris, and the foreign material, so that wound healing can proceed [Rinder & Fitch 1996; Williams & Solomkin 1999; Wagner & Roth 2000]. Although

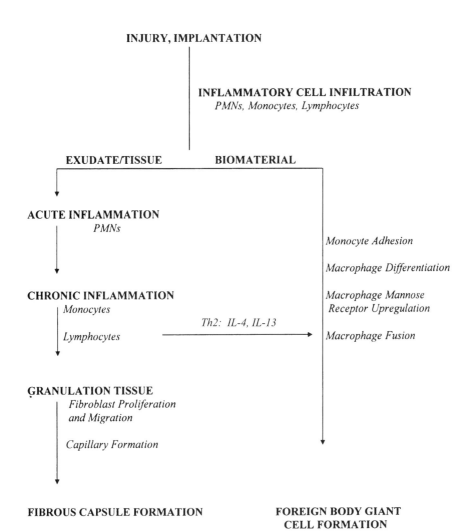

INJURY, IMPLANTATION

INFLAMMATORY CELL INFILTRATION
PMNs, Monocytes, Lymphocytes

EXUDATE/TISSUE   BIOMATERIAL

ACUTE INFLAMMATION
PMNs

Monocyte Adhesion

Macrophage Differentiation

CHRONIC INFLAMMATION   Macrophage Mannose
Monocytes   Receptor Upregulation

Lymphocytes   *Th2: IL-4, IL-13*   Macrophage Fusion

GRANULATION TISSUE
*Fibroblast Proliferation
and Migration*

Capillary Formation

FIBROUS CAPSULE FORMATION   FOREIGN BODY GIANT
                                   CELL FORMATION

Figure 2.25. Tissue or cellular host response to injury. The presence of the injury and the biomaterial induce an inflammatory and wound healing response that successively progresses through the following phases: acute inflammation, chronic inflammation, granulation tissue, and fibrous capsule formation. During this process, macrophages and foreign body giant cells undergo "frustrated phagocytosis" in an attempt to break down the device (courtesy of James M. Anderson, MD, PhD).

implanted biomaterials are not generally phagocytosed by PMNs or macrophages because of the disparity in size, certain events in phagocytosis are known to occur. While the implant size may prevent its total ingestion by an individual PMN or macrophage, they will attach to the device and undergo what is termed *frustrated phagocytosis* [Anderson 1996]. This process does not involve engulfment of the biomaterial but does cause the extracellular release of leukocyte products (lysosomal enzymes,

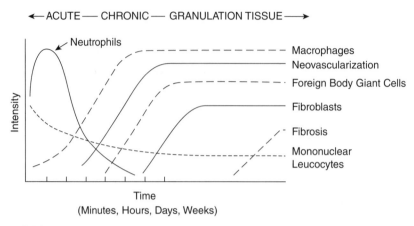

Figure 2.26. The temporal modulation of the cellularity present in the various stages of the inflammatory and wound healing response. The intensity and time variables are dependent upon the extent of the injury created by implantation and the size, shape, topography, and chemical/physical properties of the biomaterial (courtesy of James M. Anderson, MD, PhD).

proteases, and free radicals) in an attempt to degrade the biomaterial [Anderson 1993] (Figure 2.27). In general, the number of PMNs throughout the implantation time is an indicator of material–tissue compatibility or even toxic effect induced by the biomaterial leading to unsatisfactory overall *biocompatibility* (the impact—mechanical, chemical, biological—of the implanted biomaterial on the body) [Anderson 1996; Rinder & Fitch 1996]. A steady PMN population presence over time suggests continued cellular migration from the vascular system since the lifetime of the PMN is relatively short (48 hours).

Adsorption of plasma proteins onto the implant surface is considered to be the initial event in tissue–biomaterial interactions [Courtney et al. 1994; Hallab et al. 1995; Anderson 1996; Hoffman 1999; Werner & Jacobasch 1999; Gray 2004; Ratner & Bryant 2004]. The initial contact between a biomaterial surface and blood proteins results in the coating of the surface within seconds to minutes followed by competitive exchange between different proteins such as albumin, immunoglobulins (IgGs), and fibrinogen, which are reported to adsorb at high concentrations on the surface of any implanted foreign body, such as a biomaterial [Gray 2004; Ratner & Bryant 2004]. It has been suggested that spontaneous adsorption of fibrinogen is instrumental in the initiation of the acute inflammatory response [Tang & Eaton 1993]. Specifically, fibrinogen adsorption onto biomaterials is followed by a conformational modification of the quaternary structure of the adsorbed proteins, with the exposure of epitopes P_1 and P_2 responsible for interacting with phagocyte integrin Mac-1 [Hu et al. 2001]. Such process supports the accepted cause–effect relationship between protein adsorption and aggressive cellular adhesion, leading to enhanced *biofouling* (the effect of the body on the overall performance of the biomaterial) [Wisniewski et al. 2000].

The process of protein adsorption is controlled by the characteristics of the biomaterial in contact with the biological environment. Such determining features include

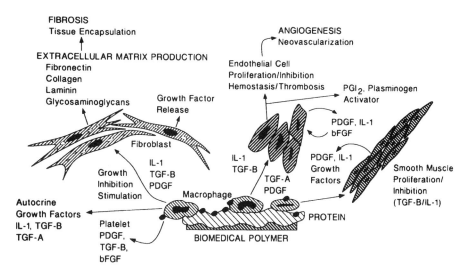

Figure 2.27. The initial protein adsorption is followed by cellular adhesion onto the protein-coated surface of the device. Adherent macrophages are responsible for the release of a host of factors involved in the early and late stages of the inflammatory and wound healing response. In addition, they play a significant role in the process of "frustrated phagocytosis" (courtesy of James M. Anderson, MD, PhD).

topography, charge density, distribution and mobility, surface groups (chain length, hydrophobicity, and hydrophilicity), structural ordering (soft to hard segment ratio and distribution, amorphous domains, and polymer chain mobility), and the extent of hydration [Schlosser & Ziegler 1997]. Therefore, it is believed that the adsorption of a single protein molecule involves several types of binding sites resulting in various strengths of binding for different proteins. Consequently, early adsorbed proteins could be desorbed and exchanged with other proteins on the surface of the biomaterial, never truly reaching a steady state in protein adsorption [Schlosser & Ziegler 1997].

The protein deposition is a dynamic process during which proteins undergo a continuous process of adsorption and desorption, multilayer formation, and denaturation. This process is controlled by the implant environment which is responding to the surface chemistry of the biomaterial and to its geometry (shape, size). The early phase of protein adsorption effectively alters the biomaterial surface, and it is this modified surface that is presented to the cellular elements [Kyrolainen et al. 1995]. By changing their conformation as a result of biomaterial surface adsorption, native proteins may be recognized by the host immune system as nonself, therefore enhancing the inflammatory response beyond the levels required to address an injury, in the absence of an implanted biomaterial. The protein adsorption could lead to aggressive cellular adhesion.

The presence of inflammatory stimuli leads to *chronic inflammation*. The chronic inflammatory response to biomaterials is usually of short duration and is confined to

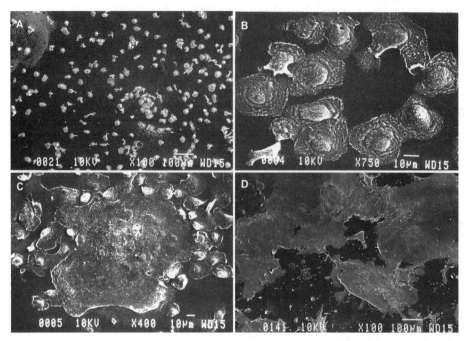

Figure 2.28. The role of macrophages is to phagocytose foreign bodies, in this case, a biomaterial. The disparity in size between these cells and the implanted biomaterial leads to the process of frustrated phagocytosis by the foreign body giant cells (FBGC). The tissue macrophages (A) migrate toward each other (B), and their cytoplasm fuses to FBGC. The removal of FBGC from the surface of a biomaterial upon explantation reveals surface degradation (D) as a result of the frustrated phagocytosis. Limited surface degradation indicates better material biocompatibility (courtesy of James M. Anderson, MD, PhD).

the implant site. Mononuclear cells, primarily macrophages, monocytes, and lymphocytes are all involved in this response, as well as the proliferation of blood vessels, fibroblasts, and connective tissue [Anderson 1996] (Figure 2.26). Macrophages are by far the most significant cells in determining the biocompatibility of implanted materials [Anderson 1996; Collier & Anderson 2002] (Figure 2.28).

Blood monocytes migrate to the implant site, differentiate into macrophages, which can adhere to the biomaterial and can become activated [Anderson 1996; Kao 1999; Janatova 2000; Hamilton 2003] (Figure 2.29). It has been shown that an activated macrophage is capable of producing and secreting important bioactive agents such as chemotactic factors, reactive oxygen metabolites, complement components and cytokines, coagulation factors, among others [Anderson 1996; Janatova 2000]. These agents have the potential to degrade or compromise the material integrity, as well as to regulate the wound healing response [Anderson 1996]. Adherent macrophages have also been shown to fuse and form FBGCs in vivo, in response to a large nonphagocytosable surface of a biomaterial [Anderson 2000] (Figure 2.29). It is believed that FBGCs retain

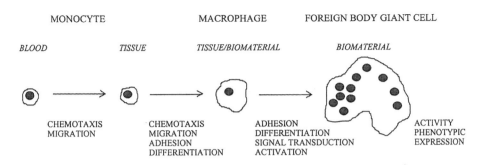

MONOCYTE   MACROPHAGE   FOREIGN BODY GIANT CELL

BLOOD   *TISSUE*   *TISSUE/BIOMATERIAL*   *BIOMATERIAL*

CHEMOTAXIS   CHEMOTAXIS   ADHESION   ACTIVITY
MIGRATION   MIGRATION   DIFFERENTIATION   PHENOTYPIC
                   ADHESION   SIGNAL TRANSDUCTION   EXPRESSION
                   DIFFERENTIATION   ACTIVATION

Figure 2.29. The inflammatory and wound healing response induced by the presence of an injury as a result of biomaterial implantation initiates the migration of blood monocytes from the ruptured vessels to the site of implantation where they become tissue macrophages. Through cellular fusion, tissue macrophages that had adhered onto the surface of the biomaterial undergo a process of fusion, thus forming foreign body giant cells in an attempt to digest the foreign body (courtesy of James M. Anderson, MD, PhD).

many of the biochemical properties of macrophages [Anderson 1996]. It has been reported that the microenvironment at the macrophage or FBGC/material surface interface is characterized by a decline in the pH, compared to that of the surrounding environment [Heming et al. 2001].

Regardless of the implantation site, intravascular or subcutaneous, the blood–biomaterial interaction is responsible for the formation of microthrombi and activation of complement factors as well as phagocytic cells [Kao 1999; Hamilton 2003]. The presence of the microthrombi, along with that of the phagocytic cells, is responsible for the generation of a local environment that distorts the normal (uninjured) biochemistry [Schlosser & Ziegler 1997].

Granulation tissue is the hallmark of healing inflammation and actually begins within 24 hours of the implantation injury [Anderson 1996]. At this point, fibroblasts and vascular endothelial cells proliferate at the implant site and begin to form a histologically pink granular tissue (H&E stain) which is called granulation tissue [Anderson 1996]. The granulation tissue displays neovascularization and a large number of proliferating fibroblasts. Macrophages are also present within granulation tissue and are considered to be central to the coordination of reorganization and reparative events due to their release of additional bioactive agents such as growth factors and cytokines [Anderson 1996; Janatova 2000]. Angiogenesis occurs by budding of preexisting vessels, where new endothelial cells organize themselves into capillary tubes. The fibroblasts are active in synthesizing proteoglycans and collagen [Janatova 2000].

Granulation tissue and the formation of the FBGCs characterize the next stage, which is known as the *foreign body reaction*. This is considered a normal wound healing response to "inert" biomaterials, and the absence of mononuclear cells, including lymphocytes and plasma cells, indicates the resolution of the chronic phase [Anderson 1996; Kirkpatrick et al. 2002; Voskerician et al. 2003]. The observed foreign body reaction can be altered by the implant geometry and surface topography. The number

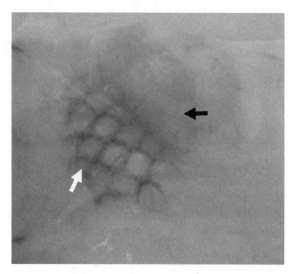

Figure 2.30. Fibrous encapsulation of woven polypropylene surgical mesh (white arrow). The biocompatibility of the surgical mesh was tested in the subcutaneous tissue of the rat. At 21 days postimplantation, a fine transparent fibrous capsule encapsulated the mesh. During excision of the tissue for analysis, underlying muscle was also harvested in an attempt to preserve intact the fragile fibrous capsule (black arrow).

of macrophages, FBGCs, fibroblasts, and new capillaries will vary depending on surface characteristics such as smoothness, or conversely, roughness [Anderson 1996]. It is probable that this reaction consisting of macrophages and FBGCs will persist at the tissue–material interface for extended periods of time (months to years) [Anderson 1996].

Once the local tissue inflammation response has determined that the biomaterial cannot be ingested or expelled, it is effectively walled off from the local tissue environment by complete *fibrous encapsulation* [Anderson 1996; Shawgo et al. 2004; Voskerician 2004a; Voskerician et al. 2004b] (Figure 2.30). The thickness of the fibrous capsule around the material has been used as a measure of the material-induced biocompatibility [Anderson 1996; Voskerician et al. 2003; Shawgo et al. 2004; Voskerician et al. 2004b]. The composition and aspect of this capsule is determined mainly by the extent of macrophage presence in response to particulates or other agents originating from the material. Relative motion of the implanted biomaterial against the surrounding tissue represents another aspect influencing the macrophages to produce fibrogenic agents that result in thicker capsules [Goodman 1994; Kirkpatrick et al. 2002]. Type III collagen is predominant especially in the early stages of wound healing, and its synthesis by fibroblasts is increased as during the process of fibrous encapsulation [Anderson 1996; Chavrier & Couble 1999]. However, with time, the amount of type III collagen decreases as it is replaced by type I collagen, which is the primary collagen composition surrounding the implanted biomaterials [Anderson 1996; Chavrier & Couble 1999; Voskerician et al. 2004b] (Figure 2.31). Over extended periods of implantation, local

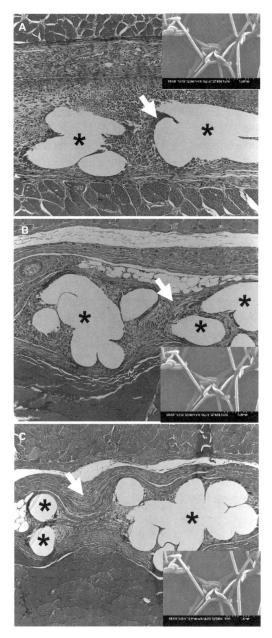

Figure 2.31. The process of wound healing in the presence of a biomaterial (Masson's Tri-Chrome Blue). The biocompatibility of a woven material was evaluated subcutaneously in rat at (A) 1 week, (B) 6 weeks, and (C) 12 weeks postimplantation. The initial inflammatory response at 1 week postimplantation (A, white arrow) diminished over time leading to collagen deposition (B, C, white arrow). In this case, the architecture of the biomaterial allows for collagen encapsulation of the polymer fibers and deposition of collagen interwoven throughout the open spaces of the biomaterial. The location of the biomaterial is identified by an asterisk (lost during histology processing). Scanning electron microscopy of the biomaterial has been placed as inset within each image. Masson's Tri-Chrome Blue stain was used to clearly identify the presence of collagen (blue).

hypoxia and decreased pH may adversely affect the intended performance of the biomaterial through alteration of its original mechanical properties and surface chemistries
[Kyrolainen et al. 1995]. Fibrous capsule formation around the biomaterial may in fact
isolate it from the surrounding biological environment, thus affecting its intended use
[Kyrolainen et al. 1995; Voskerician et al. 2004a].

Biomaterials, individual or in the form of medical devices, have been adopted by the
medical community in addressing a plethora of clinical challenges. Spanning from a
Band-Aid to implantable defibrillators, biomaterials provide patients with second chances,
from healing a skin cut with minimal scar tissue formation to extending human life!

EXERCISES/QUESTIONS FOR CHAPTER 2

1. What are the factors used in the classification of epithelial tissue?
2. What are the major functions of the skin?
3. Describe the process of cellular keratinization.
4. What is the pathophysiology of the diabetic foot ulcers (DFUs)?
5. What are the cytoarchitectural and contractile differences between the three different types of muscle tissue?
6. Why is muscular dystrophy such a devastating disease?
7. Describe the process of restenosis.
8. What are the basic functions of the resident cells of the connective tissue proper?
9. Justify the high incidence of glycine amino acid residues present in the structure
   of the collagen triple helix.
10. What are the most notable types of collagen present in the human body and to what
    tissues do they belong?
11. What are the fundamental distinctions between the three different types of
    cartilage?
12. Describe the critical roles of macrophages in the context of inflammation and
    wound healing.
13. Describe the successive stages of the inflammatory and wound healing response in
    the presence of an implanted biomaterial.

REFERENCES

Abbas AK, Lichtman AH, Pober JS (2000) Cellular and Molecular Immunology. Philadelphia:
    W.B. Saunders Company.
Anandi C, Alaguraja D, Natarajan V, Ramanathan M, Subramaniam CS, Thulasiram M, Sumithra
    S (2004) Bacteriology of diabetic foot lesions. Indian J Med Microbiol 22(3): 175–178.
Anderson J (1996) Inflammation, wound healing, and the foreign body response. In: An Introduc-
    tion to Materials in Medicine, ed. BD Ratner, 165–173. San Diego, CA: Academic Press.
Anderson JM (1993) Mechanisms of inflammation and infection with implanted devices. Car-
    diovasc Pathol 2(3): 33S–41S.

Anderson JM (2000) Multinucleated giant cells. Curr Opin Hematol 7: 40–47.

Anderson JM (2001) Biological responses to materials. Annu Rev Mater Res 31: 81–110.

Andreadis ST (2007) Gene-modified tissue-engineered skin: the next generation of skin substitutes. Adv Biochem Eng Biotechnol 103: 241–274.

Armstrong DG, Lavery LA (1998) Diabetic foot ulcers: prevention, diagnosis and classification. Am Fam Physician 57(6): 1325–1332, 1337–1338.

Aschner P (2002) Diabetes trends in Latin America. Diabetes Metab Res Rev 18(Suppl. 3): S27–S31.

Ban K, Ueki T, Tamada Y, Saito T, Imabayashi S, Watanabe M (2003) Electrical communication between glucose oxidase and electrodes mediated by phenothiazine-labeled poly(ethylene oxide) bonded to lysine residues on the enzyme surface. Anal Chem 75(4): 910–917.

Beltrami CA, Di Loreto C, Finato N, Rocco M, Artico D, Cigola E, Gambert SR, Olivetti G, Kajstura J, Anversa P (1997) Proliferating cell nuclear antigen (PCNA), DNA synthesis and mitosis in myocytes following cardiac transplantation in man. J Mol Cell Cardiol 29(10): 2789–2802.

Brem H, Tomic-Canic M (2007) Cellular and molecular basis of wound healing in diabetes. J Clin Invest 117(5): 1219–1222.

Cancedda R, Dozin B, Giannoni P, Quarto R (2003) Tissue engineering and cell therapy of cartilage and bone. Matrix Biol 22(1): 81–91.

Cantor CR, Schimmel PR (1980) Biophysical Chemistry Part I: The Conformation of Biological Macromolecules. New York: W.H. Freeman Company.

Chavrier CA, Couble ML (1999) Ultrastructural immunohistochemical study of interstitial collagenous components of the healthy human keratinized mucosa surrounding implants. Int J Oral Maxillofac Implants 14(1): 108–112.

Chen X, Matsumoto N, Hu Y, Wilson GS (2002) Electrochemically mediated electrodeposition/ electropolymerization to yield a glucose microbiosensor with improved characteristics. Anal Chem 74: 368–372.

Chen YW, Smith ML, Sheets M, Ballaron S, Trevillyan JM, Burke SE, Rosenberg T, Henry C, Wagner R, Bauch J, Marsh K, Fey TA, Hsieh G, Gauvin D, Mollison KW, Carter GW, Djuric SW (2007) Zotarolimus, a novel sirolimus analogue with potent anti-proliferative activity on coronary smooth muscle cells and reduced potential for systemic immunosuppression. J Cardiovasc Pharmacol 49(4): 228–235.

Christman KL, Lee RJ (2006) Biomaterials for the treatment of myocardial infarction. J Am Coll Cardiol 48(5): 907–913.

Clark RAF, Ghosh K, Tonnesen MG (2007) Tissue engineering for cutaneous wounds. J Invest Dermatol 127(5): 1018–1029.

Collier TO, Anderson JM (2002) Protein and surface effects on monocyte and macrophage adhesion, maturation, and survival. J Biomed Mater Res 60(3): 487–496.

Cotran RS, Kumar V, Collins T (1999) Pathologic Basis of Disease. Philadelphia: W.B. Saunders Company.

Courtney JM, Lamba NMK, Sundaram S, Forbes CD (1994) Biomaterials for blood-contacting applications. Biomaterials 15(10): 737–744.

Daemen J, Serruys PW (2007) Drug-eluting stent update 2007. Part I: A survey of current and future generation drug-eluting stents: meaningful advances or more of the same? Circulation 116(3): 316–328.

Engelmann MG, Franz WM (2006) Stem cell therapy after myocardial infarction: ready for clinical application? Curr Opin Mol Ther 8(5): 396–414.

Fu X (2005) Skin ulcers in lower extremities: the epidemiology and management in China. Int J Low Extrem Wounds 4(1): 4–6.

Gao J, Dennis JE, Solchaga LA, Goldberg VM, Caplan AI (2002) Repair of osteochondral defect with tissue-engineered two-phase composite material of injectable calcium phosphate and hyaluronan sponge. Tissue Eng 8(5): 827–837.

Gerritsen M, Kros A, Sprakel V, Lutterman JA, Nolte RJ, Jansen JA (2000) Biocompatibility evaluation of sol-gel coatings for subcutaneously implantable glucose sensors. Biomaterials 21(1): 71–78.

Goodman SB (1994) The effects of micromotion and particulate materials on tissue differentiation. Bone chamber studies in rabbits. Acta Orthop Scand Suppl 258: 1–43.

Gray JJ (2004) The interaction of proteins with solid surfaces. Curr Opin Struct Biol 14(1): 110–115.

Hallab NJ, Bundy KJ, O'Connor K, Clark R, Moses RL (1995) Cell adhesion to biomaterials: correlations between surface charge, surface roughness, adsorbed protein, and cell morphology. J Long Term Eff Med Implants 5(3): 209–231.

Hamilton JA (2003) Nondisposable materials, chronic inflammation, and adjuvant action. J Leukoc Biol 73(6): 702–712.

Heming TA, Davé SK, Tuazon DM, Chopra AK, Peterson JW, Bidani A (2001) Effects of extracellular pH on tumour necrosis factor-alpha production by resident alveolar macrophages. Clin Sci (Lond) 101(3): 267–274.

Hoffman AS (1999) Non-fouling surface technologies. J Biomater Sci Polym Ed 10(10): 1011–1014.

Hu W-J, Eaton JW, Ugarova TP, Tang L (2001) Molecular basis of biomaterial-mediated foreign body reactions. Blood 98(4): 1231–1238.

Hung CT, Lima EG, Mauck RL, Takai E, LeRoux MA, Lu HH, Stark RG, Guo XE, Ateshian GA (2003) Anatomically shaped osteochondral constructs for articular cartilage repair. J Biomech 36(12): 1853–1864.

Hunziker EB, Driesang IM (2003) Functional barrier principle for growth-factor-based articular cartilage repair. Osteoarthritis Cartilage 11(5): 320–327.

Hunziker EB, Driesang IM, Morris EA (2001) Chondrogenesis in cartilage repair is induced by members of the transforming growth factor-beta superfamily. Clin Orthop Relat Res 391(Suppl.): S171–S181.

Jackson DW, Simon TM, Aberman HM (2001) Symptomatic articular cartilage degeneration: the impact in the new millennium. Clin Orthop Relat Res 391(Suppl.): S14–S25.

Janatova J (2000) Activation and control of complement, inflammation, and infection associated with the use of biomedical polymers. ASAIO J 46(6): S53–S62.

Kao WJ (1999) Evaluation of protein-modulated macrophage behavior on biomaterials: designing biomimetic materials for cellular engineering. Biomaterials 20(23–24): 2213–2221.

Kirkpatrick CJ, Krump-Konvalinkova V, Unger RE, Bittinger F, Otto M, Peters K (2002) Tissue response and biomaterial integration: the efficacy of in vitro methods. Biomol Eng 19(2–6): 211–217.

Kubo H, Shimizu T, Yamato M, Fujimoto T, Okano T (2007) Creation of myocardial tubes using cardiomyocyte sheets and an in vitro cell sheet-wrapping device. Biomaterials 28(24): 3508–3516.

Kyrolainen M, Rigsby P, Eddy S, Vadgama P (1995) Bio-/haemocompatibility: implications and outcomes for sensors? Acta Anaesthesiol Scand Suppl 104: 55–60.

Lynn AK, Brooks RA, Bonfield W, Rushton N (2004) Repair of defects in articular joints. Prospects for material-based solutions in tissue engineering. J Bone Joint Surg Br 86(8): 1093–1099.

Miano JM, Vlasic N, Tota RR, Stemerman MB (1993) Localization of Fos and Jun proteins in rat aortic smooth muscle cells after vascular injury. Am J Pathol 142(3): 715–724.

Moussy F, Reichert WM (2000) Biomaterials community examines biosensor biocompatibility. Diabetes Technol Ther 2(3): 473–477.

Nikol S, Huehns TY, Hiifling B (1996) Molecular biology and post-angioplasty restenosis. Atherosclerosis 123(1–2): 17–31.

Pakala R, Willerson JT, Benedict CR (1997) Effect of serotonin, thromboxane A2, and specific receptor antagonists on vascular smooth muscle cell proliferation. Circulation 96(7): 2280–2286.

Palmisano F, Zambonin PG, Centonze D, Quinto M (2002) A disposable, reagentless, third-generation glucose biosensor based on overoxidized poly(pyrrole)/tetrathiafulvalene-tetracyanoquinodimethane composite. Anal Chem 74(23): 5913–5918.

Pires NMM, Eefting D, de Vries MR, Quax PHA, Jukema JW (2007) Sirolimus and paclitaxel provoke different vascular pathological responses after local delivery in a murine model for restenosis on underlying atherosclerotic arteries. Heart 93(8): 922–927.

Quinn CP, Pathak CP, Heller A, Hubbell JA (1995) Photo-crosslinked copolymers of 2-hydroxyethyl methacrylate, poly(ethylene glycol) tetra-acrylate and ethylene dimethacrylate for improving biocompatibility of biosensors. Biomaterials 16(5): 389–396.

Ratner BD (2001) Replacing and renewing: synthetic materials, biomimetics, and tissue engineering in implant dentistry. J Dent Educ 65(12): 1340–1347.

Ratner BD, Bryant SJ (2004) Biomaterials: where we have been and where we are going. Annu Rev Biomed Eng 6: 41–75.

Rinder C, Fitch J (1996) Amplification of the inflammatory response: adhesion molecules associated with platelet/white cell responses. J Cardiovasc Pharmacol 27(Suppl. 1): S6–S12.

Rodino-Klapac LR, Chicoine LG, Kaspar BK, Mendell JR (2007) Gene therapy for duchenne muscular dystrophy: expectations and challenges. Arch Neurol 64(9): 1236–1241.

Ross MH, Kaye GI, Pawlina W (2003) Histology: A Text and Atlas. Philadelphia: Lippincott Williams & Wilkins.

Schaefer D, Martin I, Jundt G, Seidel J, Heberer M, Grodzinsky A, Bergin I, Vunjak-Novakovic G, Freed LE (2002) Tissue-engineered composites for the repair of large osteochondral defects. Arthritis Rheum 46(9): 2524–2534.

Schlosser M, Ziegler M (1997) Biocompatibility of active implantable devices. Biosensors. In: The Body: Continuous In Vivo Monitoring, ed. D Fraser, 139–170. Chichester, England: John Wiley & Sons Ltd.

Shapiro F, Koide S, Glimcher MJ (1993) Cell origin and differentiation in the repair of full-thickness defects of articular cartilage. J Bone Joint Surg Am 75(4): 532–553.

Shawgo RS, Voskerician G, Duc HL, Li Y, Lynn A, MacEwan M, Langer R, Anderson JM, Cima MJ (2004) Repeated in vivo electrochemical activation and the biological effects of microelectromechanical systems drug delivery device. J Biomed Mater Res 71A(4): 559–568.

Sherwood JK, Riley SL, Palazzolo R, Brown SC, Monkhouse DC, Coates M, Griffith LG, Landeen LK, Ratcliffe A (2002) A three-dimensional osteochondral composite scaffold for articular cartilage repair. Biomaterials 23(24): 4739–4751.

Silva SY, Rueda LC, Márquez GA, López M, Smith DJ, Calderón CA, Castillo JC, Matute J, Rueda-Clausen CF, Orduz A, Silva FA, Kampeerapappun P, Bhide M, López-Jaramillo P

(2007) Double blind, randomized, placebo controlled clinical trial for the treatment of diabetic foot ulcers, using a nitric oxide releasing patch: PATHON. Trials 8: 26.

Solchaga LA, Goldberg VM, Caplan AI (2001) Cartilage regeneration using principles of tissue engineering. Clin Orthop Relat Res 391(Suppl.): S161–S170.

Stojadinovic O, Brem H, Vouthounis C, Lee B, Fallon J, Stallcup M, Merchant A, Galiano RD, Tomic-Canic M (2005) Molecular pathogenesis of chronic wounds: the role of beta-catenin and c-myc in the inhibition of epithelialization and wound healing. Am J Pathol 167(1): 59–69.

Tang L, Eaton JW (1993) Fibrin(ogen) mediates acute inflammatory responses to biomaterials. J Exp Med 178(6): 2147–2156.

Toba T, Nakamura T, Lynn AK, Matsumoto K, Fukuda S, Yoshitani M, Hori Y, Shimizu Y (2002) Evaluation of peripheral nerve regeneration across an 80-mm gap using a polyglycolic acid (PGA)—collagen nerve conduit filled with laminin-soaked collagen sponge in dogs. Int J Artif Organs 25(3): 230–237.

Voskerician G (2004) Biological Design Criteria for Implantable Electrochemical Devices. Biomedical Engineering. Cleveland, OH: Case Western Reserve University.

Voskerician G, Shive MS, Shawgo RS, von Recum H, Anderson JM, Cima MJ, Langer R (2003) Biocompatibility and biofouling of MEMS drug delivery devices. Biomaterials 24(11): 1959–1967.

Voskerician G, Liu C-C, Anderson JM (2004a) Electrochemical characterization and in vivo biocompatibility of thick film printed sensor for in vivo continuous monitoring. IEEE Sens 5(6): 1147–1158.

Voskerician G, Shawgo RS, Hiltner PA, Anderson JM, Cima MJ, Langer R (2004b) In vivo inflammatory and wound healing effects of gold electrode voltammetry for MEMS microreservoir drug delivery device. IEEE Trans Biomed Eng 51(4): 627–635.

Wagner JG, Roth RA (2000) Neutrophil migration mechanisms, with an emphasis on the pulmonary vasculature. Pharmacol Rev 52(3): 349–374.

Ward WK, Slobodzian EP, Tiekotter KL, Wood MD (2002) The effect of microgeometry, implant thickness and polyurethane chemistry on the foreign body response to subcutaneous implants. Biomaterials 23(21): 4185–4192.

Weintraub WS (2007) The pathophysiology and burden of restenosis. Am J Cardiol 100(5A): 3K–9K.

Werner C, Jacobasch HJ (1999) Surface characterization of polymers for medical devices. Int J Artif Organs 22(3): 160–176.

Wild S, Roglic G, Green A, Sicree R, King H (2004) Global prevalence of diabetes: estimates for the year 2000 and projections for 2030. Diabetes Care 27(5): 1047–1053.

Williams MA, Solomkin JS (1999) Integrin-mediated signaling in human neutrophil functioning. J Leukoc Biol 65(6): 725–736.

Wisniewski N, Moussy F, Reichert WM (2000) Characterization of implantable biosensor membrane biofouling. Fresenius J Anal Chem 366(6–7): 611–621.

Young BHJ (2000) Wheather's Functional Histology: A Text and Colour Atlas. Edinburgh, UK: Churchill Livingstone.

Yusuf S, Reddy S, Ôunpuu S, Anand S (2001) Global burden of cardiovascular diseases. Part I: General considerations, the epidemiologic transition, risk factors, and impact of urbanization. Circulation 104(22): 2746–2753.

3

HARD TISSUE STRUCTURE AND FUNCTIONALITY

Antonio Merolli and Paolo Tranquilli Leali

3.1 DEFINITION OF HARD TISSUES

Bone and cartilage have a well-defined stand-alone three-dimensional (3D) shape owing to their inner structure, which deforms very little under applied loads. This is the reason why they are cumulatively called "hard tissues" in contrast to "soft tissues," a term that encompasses structures, which, like muscle or skin, are able to deform far more greatly.

Most of this chapter will be devoted to load-bearing hard tissues such as those present in articular cartilage and those present in bones of the weight-bearing limbs.

3.2 ARTICULAR CARTILAGE

According to the definition given by the American Heritage Medical Dictionary, the articular cartilage can be defined as "the cartilage covering the articular surfaces of the bones forming a synovial joint." This type of cartilage is also called "arthrodial cartilage" as well as "diarthrodial cartilage," or "investing cartilage." This is a connective tissue that derives from the mesenchyme of the developing embryo and consists of flexible rather than elastic tissue. Its semitransparent appearance has led the scientist to give it the name "hyaline."

Biomimetic, Bioresponsive, and Bioactive Materials: An Introduction to Integrating Materials with Tissues, First Edition. Edited by Matteo Santin and Gary Phillips.
© 2012 John Wiley & Sons, Inc. Published 2012 by John Wiley & Sons, Inc.

3.2.1   Structure of the Articular Cartilage

Articular cartilage constitutes the surface of sliding joints. The basic structure of a sliding joint is made up of two articular ends surrounded by a capsule and immersed into a lubricating and nutrient fluid called the "synovial fluid."

The outer side of the capsule is made of a highly resistant fibrous tissue while the inner side is made by a carpet of cells that constitutes the "synovial membrane," which is responsible for: (1) producing the lubricating fluid, (2) providing the exchange of nutrients and waste products to and from the cartilage, (3) supporting the scavenging of bacteria and other infectious agents.

Articular cartilage bears the same load of bone. For this reason, during activities like jumping or running, anatomical sites including hips, knees, or the ankle joints, articular cartilage has to sustain loads several times greater than body weight [Mow & Ratcliffe 1993; Allan 1998; Pacifici et al. 2000].

The synovial fluid is defined as a non-Newtonian fluid since it is able to adapt and change its viscosity according to the frictional load applied. Pathological conditions, like arthritis and infection, may alter the composition and performance of the synovial fluid.

Articular cartilage and subchondral bone (literally "the bone under the cartilage") have significant anatomical and functional relationships but, at the same time, very important differences. The most notable difference is the lack of vascularization of articular cartilage. Indeed, there are no blood vessels that provide nutrients to the cartilage cells, and for this reason nutrients and waste products have to diffuse into the highly hydrated medium of the cartilage extracellular matrix, and be exchanged with the synovial fluid and ultimately with the synovial membrane.

As a matter of fact, when pathological conditions affect the articular cartilage leading to its invasion by blood vessels, the cartilage tissue is replaced by either bone or fibrotic tissue. A typical example is the exuberant bone growth observed in osteoarthrosis [Burr 1998; Karsdal et al. 2008].

From the histological point of view, there are four different zones that can be identified within the articular cartilage.

Zone 1. Called the "tangential zone," it is the outer layer of the cartilage that faces the synovial fluid and the opposite end of the joint. This zone is able to resist high shear and compressive forces. Few layers of elongated cells may be distinguished inside this zone; elongation of the cells and of the collagen fibers of the extracellular matrix follows the profile of the joint end.

Zone 2. Called the "transitional" or "intermediate zone," it shows round chondrocytes dispersed within the matrix, with collagen fibers mostly oriented in an oblique fashion.

Zone 3. Named "radial zone," it shows cells that are piled up in columns, which, like the collagen fibers inside the cellular matrix, dispose themselves at right angles to the articulating surface (hence the term "radial").

The boundary between zone 3 and zone 4 is termed the "tidemark," a tiny rim 2–5 μ in thickness, which sharply demarcates uncalcified tissue from calcified tissue (Figure 3.1).

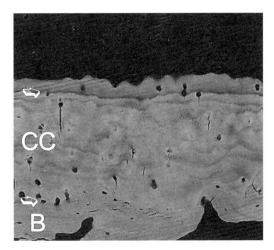

Figure 3.1. A back-scattered electron micrograph shows the boundary between zone 3 and zone 4 termed the "tidemark" (upper arrow), a tiny rim 2–5 µ in thickness. Below the "tidemark" lies the "calcified cartilage" (CC) of Zone 4 which confines (lower arrow) with subchondral bone (B).

Zone 4. Named the "calcified" or "deep zone," it interdigitates tightly with subchondral bone. Cells appear small and, in some cases, degenerated or frankly necrotic. Collagen fibers have a random orientation. While in mature bone tissue calcium salts are orderly assembled together with collagen fibers, thus achieving the mechanical performance of a composite material (see below), calcium salts in zone 4 of the articular cartilage are simply dispersed like sand washed ashore (Figure 3.2).

There are collagen fibers that actually originate from subchondral bone and enter articular cartilage. As a matter of fact, the overall arrangement of the collagen fibers in the articular cartilage has evolved as vast arcades able to withstand compressive forces.

3.2.2 Specific Mechanism Repair of the Articular Cartilage

The articular cartilage has no blood vessels and no lymphatic vessels. Chondrocytes have a great resistance to hypoxia in comparison with most of the cells in the body. The sequence of events characterizing the repair process of any tissue following damage (necrosis, inflammation, repair) does not take place in articular cartilage, and the actual response to injury is highly correlated with the severity of the damage [Ghivizzani et al. 2000]. The damages that occur in the articular cartilage can be categorized into three main types according to their severity as (1) a superficial laceration, above the tidemark, which results in a permanent injury because of the absence of inflammatory response in this totally avascular region; (2) a deep laceration progressing to the underlying subchondral bone, which will result in to a reparative fibrocartilagineous scar originating from the damage of the blood vessels of the subchondral bone; (3) a high-energy

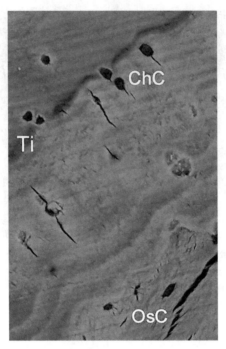

Figure 3.2. A back-scattered electron micrograph shows below the "tidemark" (Ti) the Zone 4, named "calcified or deep zone," which interdigitates tightly with subchondral bone. Chondrocytes (ChC) appear small and, in some cases, degenerated or frankly necrotic. Calcium salts are simply dispersed like sand washed ashore and do not have the orderly assembly with collagen fibers as in mature bone (OsC: osteocytes from subchondral bone).

impaction injury, which disrupts the collagen network and causes death of the chondrocytes, thus resulting in a permanent injury.

The combination of the complex arrangement of cells and fibers inside the articular cartilage (with its complex anatomical and mechanical interdigitation with subchondral bone) and with its avascularity represents a significant challenge in finding an easy clinical strategy for the repair of injuries in this type of tissue. Currently, a combined aim to provide both cells (autologous chondrocytes or stem cells) and tailor-made artificial scaffolds seeded by these cells (mostly in the form of a 3D lattice of degradable material such as polylactic acid) is the most favored approach to the problem [Lee & Shin 2007].

3.3 BONE TISSUE

According to Mosby's Medical Dictionary, bone tissue is "a hard form of connective tissue composed of osteocytes and calcified collagenous intercellular substance arranged in thin plates." Bones are the main components of the human skeleton where they play different roles in supporting and protecting tissues and organs. In addition, bones are an important reservoir for minerals, and they are involved in mineral homeostasis. Bones also host hemopoiesis, the production of blood cells by the bone marrow. Bony tissues exert their functions through a well-defined development of their structures.

3.3.1 The Structure of the Bony Tissues

During development, new bone is formed as "reticular bone," which is defined as an immature bone type where the collagen fibers are arranged in irregular random arrays and contain smaller amounts of mineral substance and a higher proportion of osteocytes (bone cells). In reticular bone, osteocytes show a polygonal morphology and are present in numbers greater than those found in mature "lamellar bone." Reticular bone is a temporary tissue that is eventually converted into lamellar bone, the latter being so-called because of the typical osteocytes arrangement in its matrix as ordered lamellae.

In the adult there are two types of bone tissue: "compact bone" (also called "cortical bone") and "trabecular bone" (also called "cancellous bone" or "spongy bone") [Burstein et al. 1975; Gibson 1985; Ascenzi 1988; Martin & Burr 1989; Martini 1989; Keaveny & Hayes 1993; Bilezikian et al. 1996; Mow & Hayes 1997; Bostrom et al. 2000; Cowin 2001; Ballock & O'Keefe 2003].

Compact bone is the outer layer of bone, and it represents approximately 80% of the skeletal system mass. Compact bone blends into trabecular bone, an inner structure that resembles a honeycomb architecture, which accounts for 20% of the bone mass. This trabecular, mesh-like bone is designed to sustain relatively high loads, and its strength is comparable to that of steel rods within a concrete structure.

Compact bone consists of closely packed "osteons" or Haversian systems (Figure 3.3a,b). The osteon consists of a central canal called the osteonic (Haversian) canal, which is surrounded by concentric rings of lamellar bone.

Between the rings, the bone cells (osteocytes) are located in spaces called "lacunae." Small channels (canaliculi) radiate from the lacunae to the osteonic (Haversian) canal. These small channels provide a passageway for the diffusion of molecules and

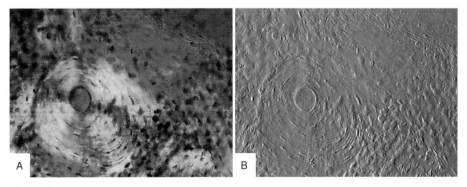

Figure 3.3. A polarized-light micrograph which shows an "osteon" or haversian system: The osteon (a) consists of a central canal surrounded by concentric rings lamellar bone (outlined by image processing in b).

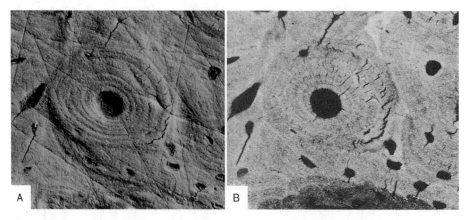

<u>Figure 3.4.</u> A scanning electron micrograph shows the concentric rings of lamellar bone of an osteon (a). Performing a back-scattered electron microscopy on the same field, it is possible to outline that bone cells (osteocytes) are located in spaces called "lacunae," and small channels (canaliculi) radiate from the lacunae to the osteonic (haversian) canal to provide passageways through the hard tissue. The osteonic canals contain a blood vessel that is parallel to the long axis of the bone and interconnects, by way of perforating canals called Volkmann's canal, with vessels on the surface of the bone (b).

biochemical signals through the hard tissue. In compact bone, the Haversian systems are packed tightly together to form what appears to be a solid mass. The boundaries between Haversian systems are called "cement lines." Each osteonic canal hosts a blood vessel that is parallel to the long axis of the bone. These blood vessels interconnect, through perforating canals called the Volkmann's canal, with vessels on the surface of the bone where the "periosteum" is present; this is a fibrous membrane that covers the outside of bone and which is rich with capillaries supporting the underlying bone, and it is populated by progenitor cells (Figure 3.4a,b).

Trabecular bone is lighter and less dense than compact bone; it consists of plates and bars of bone adjacent to small, irregular cavities that contain red bone marrow. Unlike compact bone where Harvesian canal ensure the blood supply, in trabecular bone, the canaliculi connect to the adjacent bone marrow cavities (Figure 3.5).

Although the trabeculae may appear to be arranged in a random manner, they are still able to provide maximum strength to the tissue at a level comparable to the braces that are used to support a building. The trabeculae of trabecular bone follow the lines of stress and, through remodeling, they can realign if the direction of stress changes. This realignment follows Wolff's law, which states that bone is deposited in proportion to the compression load that it has to bear (Figure 3.6).

3.3.2   The Functions of Bone Tissue

Bone tissue (or "osseous tissue") performs numerous important functions such as providing attachment sites for muscles allowing for movement of limbs; protecting vital

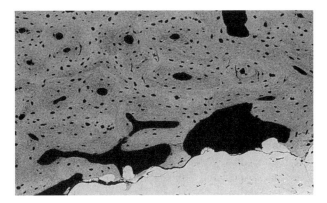

Figure 3.5. In this back-scattered electron micrograph, trabecular bone is interposed between compact bone, rich of osteonic system and rounded cells typical of very young individuals, and hydroxyapatite coating (the whitish material at the bottom of the picture).

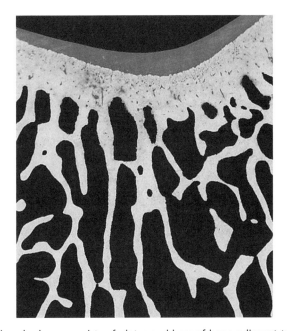

Figure 3.6. Trabecular bone consists of plates and bars of bone adjacent to small, irregular cavities that contain red bone marrow. This back-scattered electron micrograph shows trabecular bone under the intercondylar groove (the gray crescent structure on top) in a rabbit knee.

organs like the brain and heart against mechanical forces and traumas from the surrounding environment; storing calcium and phosphorus; providing the appropriate environment for hemopoiesis (the formation of blood cells); acting as a defense against acidosis; and trapping some dangerous minerals such as lead.

In addition to its well-known "mechanical" function, bone is a reservoir for minerals and has a very important "metabolic" function: It stores about 99% of the body's calcium and about 85% of the phosphorus. The storage of calcium in bone is very important to keep the blood level of calcium within a narrow range. Relatively high or low levels of calcium in blood can impair muscles, nerves, and heart functions, and even lead to their failure. In times of specific metabolic needs—for example, during pregnancy—higher levels of calcium can be mobilized from the bones and made available to the organism functions.

The role of bone in hemopoiesis is also very important; this process takes place into the marrow present inside the trabecular bone. As a matter of fact, while the marrow in the medullary cavity of an adult bone is usually of a type called "yellow marrow," which mainly functions as a fat storage tissue, the marrow in the spaces of the trabecular bone is mainly "red marrow," a type of tissue that is able to produce various types of blood cells, such as red blood cells (a process known as "erythropoiesis").

As far as the main mechanical functions are concerned, bone has to endure forces that, even in the case of light physical exercise, may account for four to five times a person's entire body weight. A constant overload subjects the bone structure to a little perceptible tear and wear, and a continuous process of reparation and replacement is constantly underway to reinforce the structure. However, the stress derived from a regular exercise supports tissue strengthening. With age, the bone structure changes, and its repair potential gradually slows down.

3.3.3   Cell Types Involved in Bone Homeostasis: The Osteoblasts and the Osteoclasts

There are two types of specialized cells that contribute to bone homeostasis: osteoblasts and osteoclasts [Lanyon 1993; Blair 1998; Roodman 1999].

Osteoblasts are bone-forming cells able to deposit extracellular matrix components (e.g., collagen) and to contribute to the formation of the tissue precursor, the osteoid (see Section 3.3.4). During this phase, calcium is withdrawn from the blood and deposited into the maturing bone extracellular matrix. Osteoblasts undergo a maturation process that leads to their differentiation from osteocytes. During the process of newly formed bone formation (a phase called "bone apposition"), osteoblasts that have become encased in the same tissue that they have produced become osteocytes. Osteocytes engage in the metabolic exchange with the blood that flows through the bones' dedicated structures (see Section 3.3.1).

Osteoblasts are found on the outer surfaces of bones and in bone cavities. A small amount of osteoblastic activity occurs continually in all living bones (on about 4% of all surfaces at any given time) to ensure a constant formation of new bone. The new bone is being laid down in successive layers of concentric circles on the inner surfaces of the cavity until it is filled by a new Haversian system. Deposition of new bone ceases when the bone reaches the surface of the blood vessels supplying the area. Therefore, the canal through which these blood vessels run is all that remains of the original cavity. This process continues until about age 40, when the activity of osteoblasts is significantly reduced and bones become more brittle.

Osteoclasts are cells devoted to bone resorption, that is, the process of the breaking down of bone structure. In an adult, bone engages in a continuous cycle of resorption and apposition that is provided by the balanced activity of both osteoblasts and osteoclasts. This cyclic process is termed "bone turnover."

Osteoclasts are generally found in small but concentrated agglomerates, and once a mass of osteoclasts begins to develop, it usually erodes the bone for about 3 weeks, forming a tunnel in the mineral phase that may be as large as 1 mm in diameter and several millimeters in length. At the end of this time, the osteoclasts disappear, and the tunnel is invaded by osteoblasts, which initiate the apposition of new bone that continues for several months.

3.3.4   Ossification, Turnover, and Remodeling

In the fetus, most of the skeleton is made up of cartilage, which is a tough, flexible connective tissue deprived of minerals or salts. As the fetus grows, osteoblasts and osteoclasts slowly replace cartilage cells, and ossification begins (Figure 3.7).

Ossification is the formation of bone by the activity of osteoblasts, which deposit a proteinaceous matrix (the osteoid tissue) and promote the addition of minerals; osteoblasts derive these minerals from blood. Calcium compounds must be present for ossification to take place. In humans, ossification is fully complete by approximately the age of 25. During this lengthening period, the stresses of physical activity induce the strengthening of the bone tissue. Bone development is also influenced by a number of factors. The most relevant factors that contribute to bone development are physical exercise, nutrition, exposure to sunlight, and hormonal secretions.

Bone turnover is the term that defines the continuous cycle of bone apposition and resorption driven by cells. Sometimes, looking at the procedures that are applied in both orthopedic surgery and traumatology, which are based on the extensive use of chisels, hammers, and screws, one may think that bone is treated like a nonliving, mineralized material (Figure 3.8). This is a misconception as bone turnover is a very significant process. It has been calculated that each year 10–20% of the whole human skeleton is being replaced by new bone. Normally, except in growing bones, the rates of bone apposition and resorption are comparable, and the total mass of bone remains constant [Burstein et al. 1976; Burr et al. 1985; Carter & Caler 1985; Odgaard & Weinans 1995; Burr 1997].

As anticipated in Section 3.3.2, Wolff's law establishes that bone is deposited in proportion to the compression load that it must carry and that trabeculae follow the lines of stress. The trabeculae can realign if the direction of stress changes, and so every bone segment at the macroscopic level constantly adapts its inner structure to external stimuli [Wolff 1986]. This process of continuous turnover and readaptation is called bone remodeling (Figure 3.9).

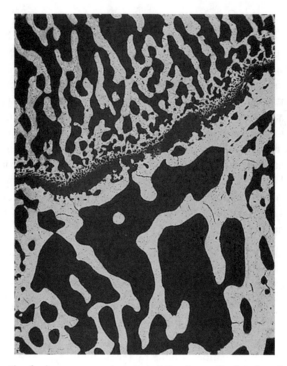

Figure 3.7. Growth of a bone segment occurs at the "growth plate" or "growth cartilage" located between the tiny trabeculae of the epiphysis and the coarser trabeculae of the diaphysis.

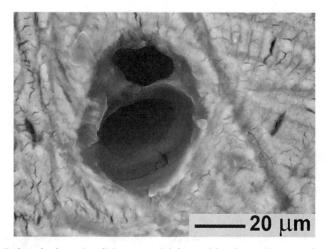

Figure 3.8. Trabecular bone is a living material; larger blood vessels can be found immersed into coarser trabeculae.

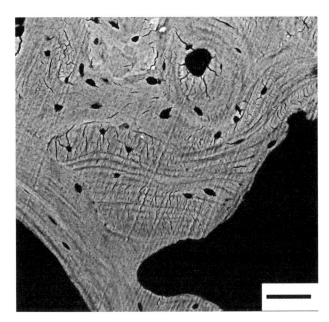

Figure 3.9. According to Wolff's law, bone is deposited in proportion to the compressional load that it must carry, and trabeculae follow the lines of stress; it constantly adapts its inner structure, so a "composition" of successive lamellar deposition can be seen, like in this back-scattered electron micrograph.

3.3.5 Bone Composite Structure and Its Effect on Mechanical Performance

Bone may be considered as an excellent composite material: it is strong but light, it can adapt to functional demands, and it can repair itself.

The organic matrix of bone is composed primarily of protein. The most abundant protein in bone is the collagen type I, which provides the tissue with flexibility; about 10% of adult bone mass is collagen. There are also a number of noncollagenous proteins that play important roles in the regulation of both cell activity and tissue biomineralization.

The main mineral component of bone is hydroxyapatite, which is an insoluble ceramic made of calcium and phosphorus; about 65% of adult bone mass is hydroxyapatite. The chemical formula for hydroxyapatite crystals is $Ca_{10}[PO_4]_6[OH]_2$.

Bone also contains small amounts of magnesium, sodium, and bicarbonate while water comprises approximately 25% of adult bone mass.

It is the unique combination of collagen fibers and hydroxyapatite that confer the appropriate mechanical properties to the bone tissue. In fact, the collagen fibers of bone have great tensile strength while the mineral phase provides the tissue with great compression strength. These combined properties are obtained through the tight chemical interactions between collagen fibers and the crystals, and provide a bony structure that has both extreme tensile and compressional strength. Despite its great compressional and tensile strength, bone does not offer a very high level of torsional strength. Indeed, bone fractures often occur as a result of torsional forces that are exerted on long bones such as those of arms and legs.

3.4   CONCLUDING REMARKS

This basic outline of the fundamental aspects of articular cartilage and bone tissue highlights the importance of cells in preserving the fundamental mechanical function of hard tissues.

In the realm of nonliving materials, this mechanical function would have been achieved mostly by adequate material properties and smart 3D arrangements. However, in living hard tissue, a stable weight-bearing structure is achieved mostly by adjusting the balance between bone resorption and bone apposition according to the direction and pace of applied loads.

Biomimetic materials for clinical applications aiming at the repair of hard tissues [Williams 1995; Inoue 2008] are successful only if they are able to achieve a satisfactory level of integration with the surrounding tissue and to preserve, or possibly restore, this dynamic cellular activity of continuous structural remodeling. It is envisaged that future biomaterials will need to exert a bioactive role and be able to participate in the remodeling process without interfering with the fine balance of biochemical and cellular processes.

EXERCISES/QUESTIONS FOR CHAPTER 3

1. Provide a definition of the articular cartilage and describe its main structural and cellular components.
2. Give examples of pathological or traumatic conditions leading to cartilage damage and indicate their most frequent origin.
3. Describe the various levels of cartilage damage and the main phases of its repair processes.
4. Which are the main bone types and what is their biomechanical function?
5. What is the osteon? Describe its main histological features.
6. Which are the main cell types in bone and their relative functions?
7. Describe the main phases of bone repair and remodelling.
8. What is the main organic component of the bone extracellular matrix?
9. List the main mineral components of the bone extracellular matrix and their role in the tissue and host homeostasis.
10. Describe the main functions of periosteum and bone marrow in long and flat bones.

REFERENCES

Allan DA (1998) Structure and physiology of joints and their relationship to repetitive strain injuries. Clin Orthop Relat Res 351: 32–38.

Ascenzi A (1988) The micromechananics versus the macromechanics of cortical bone—a comprehensive presentation. J Biomech Eng 8: 143.

Ballock RT, O'Keefe RJ (2003) Physiology and pathophysiology of the growth plate. Birth Defects Res C Embryo Today 69(2): 123–143.

Bilezikian JP, Raisz LG, Rodan GA, eds. (1996) Principles of Bone Biology. San Diego, CA: Academic Press.

Blair HC (1998) How the osteoclast degrades bone. Bioessays 20(10): 837–846.

Bostrom MPG, Boskey A, Kaufman JJ, Einhorn TA (2000) Form and function of bone. In: Orthopaedic Basic Science: Biology and Biomechanics of the Musculoskeletal System, 2nd ed., ed. JA Buckwalter, TA Einhorn, SR Simon, Rosemont, IL: AAOS Press.

Burr DB (1997) Muscle strength, bone mass and age related bone loss. J Bone Miner Res 12: 1547.

Burr DB (1998) The importance of subchondral bone in osteoarthrosis. Curr Opin Rheumatol 10(3): 256–262.

Burr DB, Martin RB, Schaffler MB, Radin EL (1985) Bone remodeling in response to in vivo fatigue damage. J Biomech 18: 189.

Burstein AH, Zika JM, Heiple KG, Klein L (1975) Contribution of collagen and mineral to the elastic-plastic properties of bone. J Bone Joint Surg A57: 956.

Burstein AH, Reilly DT, Martens M (1976) Aging of bone tissue: mechanical properties. J Bone Joint Surg 58B: 82–86.

Carter DR, Caler WE (1985) A cumulative damage model for bone fracture. J Orthop Res 3: 84–90.

Cowin SC, ed. (2001) Bone Mechanics Handbook, 2nd ed. Boca Raton, FL: CRC Press.

Ghivizzani SC, Oligino TJ, Robbins PD, Evans CH (2000) Cartilage injury and repair. Phys Med Rehabil Clin N Am 11(2): 289–307.

Gibson LJ (1985) The mechanical behaviour of cancellous bone. J Biomech 18: 317.

Inoue H (2008) The present situation and future prospects of bone substitute of bioinspired materials. Clin Calcium 18(12): 1729–1736.

Karsdal MA, Leeming DJ, Dam EB, Henriksen K, Alexandersen P, Pastoureau P, Altman RD, Christiansen C (2008) Should subchondral bone turnover be targeted when treating osteoarthritis? Osteoarthritis Cartilage 16(6): 638–646.

Keaveny TM, Hayes WC (1993) Mechanical properties of cortical and trabecular bone. In: Bone Mechanics Handbooks, Vol. 7. ed. BK Hall, 285–344. Boca Raton, FL: CRC Press.

Lanyon LE (1993) Osteocytes, strain detection, bone modeling and remodeling. Calcif Tissue Int 53(Suppl.): 102.

Lee SH, Shin H (2007) Matrices and scaffolds for delivery of bioactive molecules in bone and cartilage tissue engineering. Adv Drug Deliv Rev 59(4–5): 339–359. Review.

Martin RB, Burr DB, eds. (1989) Structure, Function, and Adaptation of Compact Bone. New York: Raven Press.

Martini FH (1989) Fundamentals of Anatomy and Physiology, 4th ed. Upper Saddle River, NJ: Prentice Hall International.

Mow VC, Hayes WC, eds. (1997) Basic Orthopaedic Biomechanics, 2nd ed. Philadelphia: Lippincott-Raven.

Mow VC, Ratcliffe A, eds. (1993) Structure and Function of Articular Cartilage. Boca Raton, FL: CRC Press.

Odgaard A, Weinans H, eds. (1995) Bone Structure and Remodeling. Singapore: World Scientific.

Pacifici M, Koyama E, Iwamoto M, Gentili C (2000) Development of articular cartilage: what do we know about it and how may it occur? Connect Tissue Res 41(3): 175–184. Review.

Roodman GD (1999) Cell biology of the osteoclast. Exp Hematol 27: 1229–1241.

Williams D (1995) Biomimetic surfaces: how man-made becomes man-like. Med Device Technol 6(1): 6–8.

Wolff J (1986) The Law of Bone Remodelling. Berlin Heidelberg: Springer-Verlag.

4

BIOMEDICAL APPLICATIONS OF BIOMIMETIC POLYMERS: THE PHOSPHORYLCHOLINE-CONTAINING POLYMERS

Andrew L. Lewis and Andrew W. Lloyd

4.1 HISTORICAL PERSPECTIVE

It was a seminal paper in *Nature* by Zwaal et al. [Zwaal et al. 1977] that first reported on the differences between the phospholipid constituents of the inside of the red blood cell membrane compared to those of the outside. It was observed that if the phospholipid bilayer is separated and the inner and outer leaflets are independently exposed to blood, the inside causes blood to clot whereas the outside does not. This begged the question as to why this should be the case and led to the conclusion that the higher proportion of negatively charged phospholipids within the inner layer was the cause of its thrombogenicity, whereras the outer layer was composed predominantly of phospholipids bearing the phosphorylcholine (PC) headgroup and was clearly hemocompatible (Figure 4.1). The potential biomedical applications of this were recognized, and a number of groups initiated research into the use of phospholipid-bearing systems to improve hemo- and biocompatibility.

Nakaya and coworkers have reported in detail on a variety of polymer systems based upon a range of phospholipid headgroups (for a detailed review of the literature from 1974–1997, see Nakaya & Li 1999). Many of the earlier systems used phospholipid diols which become incorporated within the backbone of the polymer chains using condensation polymerization methods. Although effective, the synthesis of these materials was reasonably complicated and relatively inflexible. Although Nakaya described

Biomimetic, Bioresponsive, and Bioactive Materials: An Introduction to Integrating Materials with Tissues, First Edition. Edited by Matteo Santin and Gary Phillips.
© 2012 John Wiley & Sons, Inc. Published 2012 by John Wiley & Sons, Inc.

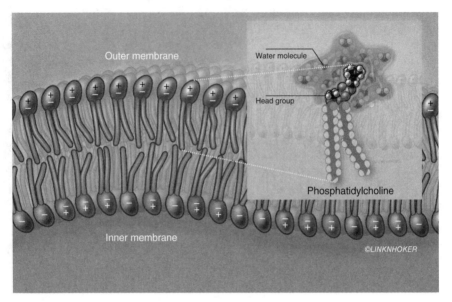

Figure 4.1. Asymmetry of the cell membrane: phospholipids in the outer layer are comprised mainly of those possessing the phosphorylcholine headgroup.

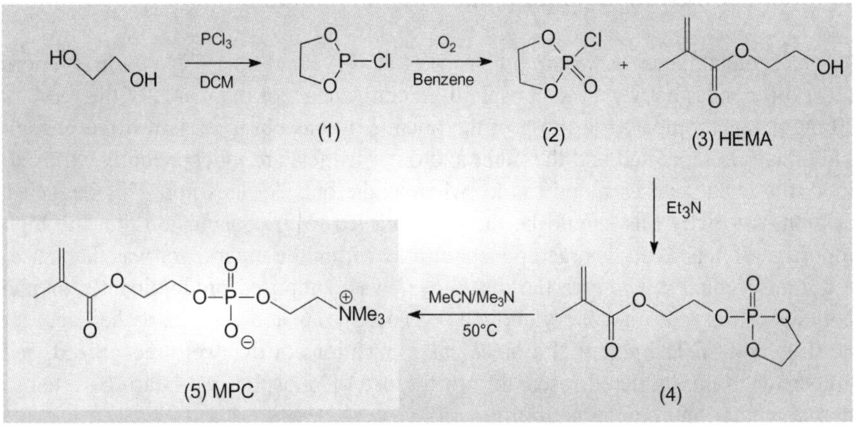

Figure 4.2. Synthesis and structure of MPC (5).

vinyl-functionalized phospholipid analogues, it was Nakabayashi and coworkers who were first to report on the synthesis of a PC-bearing methacrylate monomer known as 2-methacryloyloxyethyl phosphorylcholine (MPC) and its resulting polymers [Kadoma et al. 1978]. The monomer is prepared by the coupling of 2-hydroxyethyl methacrylate (HEMA), a commonly used material in the medical industry, with 2-chloro-2-oxo-1,3,2-dioxaphospholane, followed by a ring opening reaction of the intermediate alkoxyphospholane with trimethylamine (Figure 4.2).

Initial reports, however, were counterintuitive, suggesting the polymers were hemolytic in nature, an observation later attributed to the poor purity of the MPC monomer that had been synthesized.

The early work of Dennis Chapman's group evaluated a number of attachment and polymerization technologies to immobilize phospholipid structures onto surfaces, including for instance PC-based polyester and polyurethane (PU) systems [Hayward & Chapman 1984; Durrani et al. 1986; Hayward et al. 1986a,b; Bird et al. 1989; Hall et al. 1989]. One successful route to development of a phospholipid-based system was by engineering polymerizable diacetylenic groups in the alkyl chain portions of the phospholipid dipalmitoylphosphorylcholine (DPPC). This allowed diacetylenic phosphatidylcholine (DAPC) to be coated onto a surface and then cross-linked via the acetylene groups by use of UV radiation. These studies provided Chapman with the proof of concept he required and he formed the company "Biocompatibles" in order to develop these systems for use on a variety of medical devices. Ultimately, the focus fell on the use of MPC, aided by the development of an improved synthetic procedure that enabled the monomer to be made in high purity and good yield. The methacrylate-based chemistry opened the door to a myriad of copolymer formulations that could be tailored toward a particular end application, which have been the subject of broad patent filings by Nakabayashi in Japan and Biocompatibles in the rest of the world.

4.2 SYNTHESIS OF PC-CONTAINING POLYMERS

The homopolymerization of MPC can be achieved by solution free radical polymerization using standard techniques. Polymerization in water is possible and is accelerated in the presence of salt ions, which interact with the propagating polymer radicals and/or the MPC monomer [Wang et al. 2004]. Curiously, although water is an excellent solvent for both monomer and polymer, MPC is seen to autopolymerize when water alone is used as a solvent, leading to loss of control over the polymerization and ultimately the formation of a highly swollen crosslinked gel. It is therefore preferred to polymerize the monomer in mixtures of water with other protic solvents such as alcohols, or indeed in alcohols alone [Ma et al. 2002]. Various comonomers can be included in the polymerization mix, but the polymerization solvent system has to be adjusted to account for the solubility of both the comonomer and the resulting copolymer. Much of the early reported work on copolymer preparation used standard Schlenk techniques but failed to account for differences in both solubility and reactivity ratio of the comonomers; this led to polymers with compositions some way off that targeted, or reduced molecular weights due to premature precipitation from solution. Methods to overcome this have been reported, including polymerization under monomer-starved conditions [Lewis et al. 2000] or appropriate selection of comonomer reactivity ratios to ensure interconversion of all components in the mix [Lewis et al. 2001a]. More recently, there have been a number of reports on the use of living polymerization methods such as atom transfer radical polymerization (ATRP) [Lobb et al. 2001; Ma et al. 2002; Feng et al. 2006] or radical addition fragmentation transfer (RAFT) [Stenzel et al. 2004; Yusa et al. 2005; Iwasaki et al. 2007c], which appear to be very facile for use with MPC and

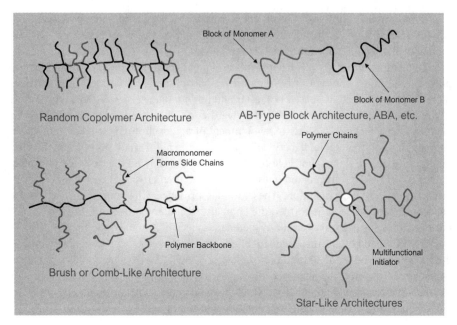

Figure 4.3. Some of the MPC polymer architectures possible using controlled radical polymerizations such as ATRP.

allow the generation of a wide variety of new polymer architectures previously unavailable by conventional polymerization techniques (Figure 4.3).

An alternative method for functionalizing a substrate with a PC polymer is to graft polymerize the MPC monomer directly from the surface. This has been described using a number of different grafting techniques, including the use of ceric ions to raise radicals at the surface [Korematsu et al. 2002], photopolymerization initiated by benzophenone [Goda et al. 2007, 2008], azo groups [Yokoyama et al. 2006], plasma-induced polymerization [Hsiue et al. 1998], and indeed ATRP using surface-bound initator systems [Feng et al. 2006; Hoven et al. 2007; Ishihara & Lifeng 2008]. Others have prepared PC polymers containing a photosensitive azide-based monomer component that allows controlled synthesis of the polymer and then subsequent graft immobilization onto the substrate by activation of the azide by light [Konno et al. 2005]. The ability to copolymerize the MPC monomer with a range of comonomers has led to the synthesis of a wide range of PC-containing polymers with interesting physicochemical and biological properties.

4.3   PHYSICOCHEMICAL PROPERTIES OF PC-CONTAINING POLYMERS

4.3.1   Antifouling Mechanisms of Action

The antithrombotic and nonfouling nature of PC polymers has been the subject of great debate and study over many years. Since the first reports of these polymers and their properties in Japan in the late 1970s, researchers have attempted to uncover the mechanism of why these materials have such good antithrombotic properties. The polymers, by design, are among the earliest examples of a biomimetic system involving a synthetic polymer created using the chemical structures found in nature. There is a plethora of published literature detailing the antifouling properties of PC polymers, demonstrating how these materials resist the adsorption of proteins when placed in contact with blood and consequently diminish the ensuing events of the clotting cascade that would ordinarily result in thrombus formation on a foreign surface (see Section 4.5.1). Moreover, significant reductions in the adhesion of many different types of cells and bacteria have also been reported, as protein adsorption is also a prerequisite for initiation of the adhesion process.

More recently, some of the work carried out in Japan on these materials was reviewed in context of the in vitro, ex vivo, and in vivo work carried out to demonstrate their nonthrombogenicity [Nakabayashi & Williams 2003]. Early work [Ueda et al. 1991] demonstrated that one mechanism for the reduced thrombogenicity was that PC polymers encourage the organization of plasma lipids into bilayered structures on the polymer surface and hence passivating it from interaction with blood proteins. A schematic of the proposed mechanism is described in Figure 4.4. This phenomenon, however, should not be considered the sole mechanism responsible for the intrinsic biocompatibility of these polymers as there are numerous examples in the literature demonstrating similar effects using lipid-free systems. A series of studies on the water-structuring abilities of the polymers have been carried out over the last few years and have sought to answer the question regarding protein resistance and hemocompatibility. Ishihara et al. [1998] utilized differential scanning calorimetry and circular dichroism spectroscopy; they demonstrated that PC polymer hydrogels were different to conventional hydrogel materials in that they possess a high free water fraction that allows proteins to come into contact with the materials, without inducing any change in their conformation. These results were later supported by Raman spectroscopic studies [Kitano et al. 2000, 2006], infrared (IR) spectroscopy [Kitano et al. 2003; 2005], and

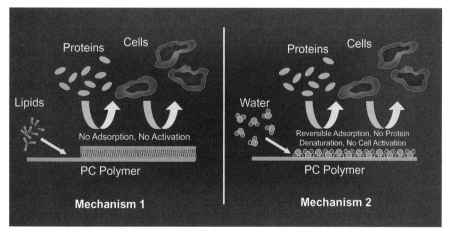

Figure 4.4. Antifouling mechanisms of action for MPC polymers.

NMR spin–spin relaxation time studies [Morisaku et al. 2005]. These essentially concluded that the zwitterionic groups of the PC hydrogel polymers, while possessing large hydration shells of free water, do not disturb the hydrogen bonding between water molecules to the extent of other charged or neutral hydrogel systems. This property, coupled with the high mobility of the pendant PC side chains in the polymers [Parker et al. 2005], allows proteins to interact reversibly with coatings of these materials without denaturing and hence without inducing the blood clotting cascade that results in thrombus formation on the surface of the material.

4.3.2   Swelling Phenomena and Structural Aspects of PC Coatings

PC polymer coatings are amphiphilic in nature due to the presence of highly polar PC moieties and hydrophobic alkyl chain components. The materials have been easily applied to surfaces of a wide variety of substrates [Campbell et al. 1994] using any of the conventional techniques such as dip, spin, or spray coating. Upon evaporation of the solvent, the polymer is deposited from solution and forms a continuous film, the thickness and integrity of which is dependent upon a number of factors including polymer molecular weight, solution concentration, coating method, and solvent. As the film forms, the polymer will orientate its comonomer components according to the nature of the environment in which it is present. In air of normal humidity, for instance, the hydrophobic groups orient at the air interface in order to minimize the interfacial free energy. This has been demonstrated using X-ray photoelectron spectroscopy (XPS) and time-of-flight–secondary ion mass spectrometry (TOF-SIMS) which shows that the PC groups are buried within the film when exposed to air or high vacuum conditions [Clarke et al. 2000] (Figure 4.5a,b). When exposed to an aqueous environment, the coating rapidly hydrates and the surface rearranges so that the PC groups are expressed at the surface (Figure 4.5d), a phenomenon demonstrated in a number of studies using contact angle measurement (Figure 4.5c). If the coating is applied from a solvent system containing some water, upon evaporation of the solvent, a small amount of water remains associated with the PC groups, and the surface retains its hydrophilic character [Lewis et al. 2000]. These properties are important when considering the intended application of the coating, as a spontaneously wettable surface may be desired and requires re-engineering of the PC polymer to avoid the hysteresis effect [Lewis et al. 2003; Yang et al. 2008].

Spectroscopic ellipsometry and neutron diffraction studies have been performed on MPC-co-lauryl (dodecyl) methacrylate (LMA) coatings, which have been shown to swell very rapidly when exposed to water, with ~50% water absorbed within the first minute [Tang et al. 2001, 2002, 2003]. The kinetics of swelling and distribution of water within the coating is dependent upon whether it is a simple physisorbed or crosslinked film. The balance of hydrophilic:hydrophobic groups and degree of cross-linking of the polymer network dominate the swelling properties. In crosslinked systems, the swelling process is biphasic, involving a rapid, essentially diffusion-controlled ingress of water, followed by a much slower phase controlled by relaxation of the polymer chains. It has been shown that although these materials are hydrogels, the large proportion of alkyl component aggregates within the structure to form hydrophobic domains.

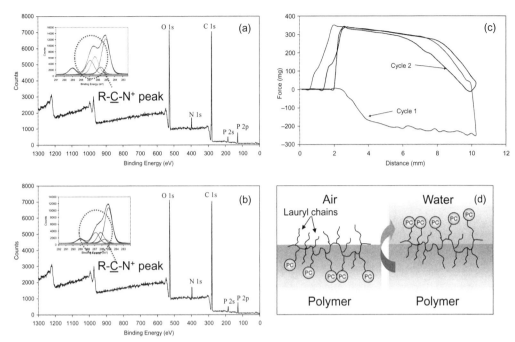

Figure 4.5. Surface mobility of the MPC group: XPS spectra under vacuum at (a) 45° and (b) 10° take-off angle demonstrating the PC group is enriched beneath the surface; (c) dynamic contact angle analysis showing hysteresis effect as the initially hydrophobic surface switches to a hydrophilic surface after exposure to water; (d) schematic representation of the surface structure in air and water.

The formation of these domains is influenced by polymer chain flexibility; the provision of energy in the form of heat or radiation during cross-linking enables the T_g to be exceeded and allows the chains to flow, facilitating the crosslinking process and hydrophobic domain formation. These properties of the PC coatings are exceedingly useful from a drug delivery perspective and are the basis of the various drug eluting stents that have been developed using this technology (see Sections 4.6.1.1 and 4.6.7.1).

A curious observation is the behavior of the polymers in certain mixed solvent systems. This was first described for solutions of PC polymers in mixtures of water and alcohol [Lewis et al. 2000] and is most obviously demonstrated using the MPC homopolymer. Although independently soluble in either water or ethanol, the polymer is seen to be insoluble in certain ratios of a mixture of the two solvents. It is believed this is a result of competition between water and alcohol around the PC headgroup and the formation of hydrated PC domains that essentially promotes transient cross-linking of the polymer (Figure 4.6). This was extended to crosslinked PC polymer hydrogels by Ishihara who observed similar strange swelling behavior in water alcohol mixtures and proposed application in drug delivery systems [Kiritoshi & Ishihara 2002]. More recently, this phenomenon has been described in terms of water–alcohol hydrogel

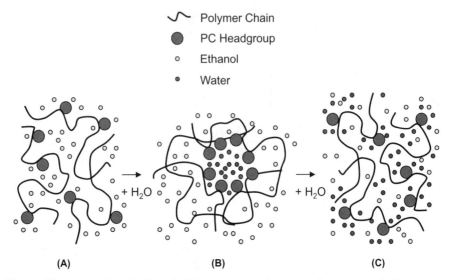

Figure 4.6. Proposed mechanism of MPC polymer gelation in certain water:alcohol mixtures.

bonding causing MPC phase separation to occur in an upper critical solution temperature (UCST)-type cononsolvency behavior [Matsuda et al. 2008].

As highly hydrated materials, the optimization of the material stability and mechanical properties can present particular challenges in the development of these materials for use in biomedical applications.

4.4   STABILITY AND MECHANICAL PROPERTY CONSIDERATIONS

4.4.1   PC Coatings and Surface Treatments

Much of the reported literature on PC polymers has focused on its application and performance as an antifouling coating material. Consequently, the stability of a PC coating on a given substrate is an important feature of its properties, unless the application requires only its transient presence for effect. Probably the most reported systems are the copolymers of MPC with butyl methacrylate (BMA) favored by the Japanese workers and collaborators and the copolymers with LMA favored by the U.K. group. These simple copolymer systems rely upon the amorphous, rubbery aliphatic comonomer component to plasticize the polymer and aid in film formation, while providing significant potential for hydrophobic interaction with such surfaces as stainless steel, a wide variety of biomedical plastics, and glass for instance. The balance of the hydrophobic portion to MPC is quite critical, as too much hydrophobe and the antifouling properties suffer, whereas too little hydrophobe and the coating is not firmly anchored and could ultimately dissolve away. This has been evident in some studies using the BMA system, as the shorter alkyl chain is inherently less adhesive compared to the LMA counterpart [Zhang et al. 1996].

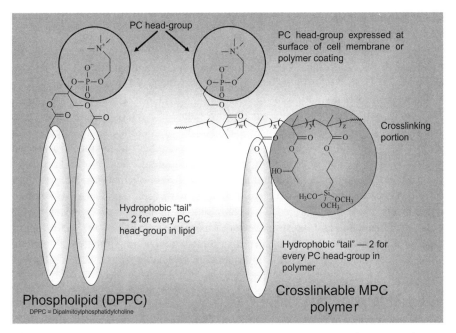

Figure 4.7. Structure of DPPC and a structurally analogous crosslinkable MPC polymer.

   The stability of the MPC-*co*-LMA physisorbed system was evaluated in studies where part of the alkyl chain of the hydrophobic portion was radiolabeled with tritium and the resulting polymer coated onto a model substrate. The film was shown to be stable as there was no detected leaching of radioactivity from the surface [Russell et al. 1993]. Nevertheless, more robust PC coatings were desired for use in applications where the coating was required to undergo significant mechanical stresses (i.e., on a stent surface which is required to expand in use). This led to the development of a polymer containing, in addition to the MPC and LMA, 3-trimethoxysilylpropyl meth-acrylate and 2-hydroxypropyl methacrylate comonomer components [Lewis et al. 2001a] (Figure 4.7). This allowed for the preparation of a linear polymer that could be subsequently solubilized in a suitable solvent, applied to a substrate by a variety of coating methods and then cured into a crosslinked network by application of heat and/ or gamma irradiation. Atomic force microscopy (AFM) techniques were used to dem-onstrate the increased force required to remove the crosslinked PC coating from the substrate compared to the physisorbed systems and how this force varied with curing and sterilization regimes [Lewis et al. 2001a]. The low level of cross-linking in this system imparts on the coating a slightly elastic property which allows it to recover following deformation; a particularly useful property with respect to expandable devices such as stents. The stability of the crosslinked PC coating on a coronary stent has demonstrated in vivo on explants removed from rabbit iliac and porcine coronary models and on a stent removed from a human subject [Lewis et al. 2002a,b, 2004b]. These studies showed the coating was still present, required the same force to remove, and had the same thickness, even after 6 months implantation; the rabbit iliac implant included the presence of a drug within the coating [Lewis et al. 2004b].

Surface modification of useful biomedical polymers such as poly(dimethylsiloxane) (PDMS) with PC groups has been demonstrated more recently by using ATRP to synthesize MPC-PDMS ABA triblock copolymers. Swelling and deswelling of PDMS with an ethanol:chloroform solution of the triblock gave rise to a stable surface expression of PC groups with improved protein and cell-resistant characteristics [Seo et al. 2008]. An alternative approach to the use of a separate coating is modification by direct grafting a PC polymer onto a substrate surface (as briefly described in Section 4.2). Surface-initiated ATRP has been used to create super-hydrophilic high density MPC brushes on silicon wafers which possess an extremely low coefficient of friction [Kobayashi et al. 2007]. Similar reductions in coefficient of friction have been reported for UV surface-grafted layers of MPC on Parylene C, a common coating applied to devices because of its electrical insulation and moisture barrier properties [Goda et al. 2007]. This lubricity property, coupled with the biological compatibility, could be important for a number of biomedical uses, including their use in orthopedic applications (see Section 4.6.4).

4.4.2   Bulk Hydrogels and Blends

Devices can be directly formed by bulk polymerization processes and conventionally cross-linked by use of common bifunctional methacrylate cross-linkers. These bulk materials hydrate to attain an equilibrium water content determined by the compositional ratio of MPC, any hydrophobic comonomer and crosslinker content. In the case of a device such as a contact lens, it is required that a balance be struck between sufficient mechanical properties to allow handling of the lens and the high water content for optical transparency, high oxygen and nutrient transport, comfort and tear film compatibility, and low propensity for protein and lipid spoilation. The Proclear® family of contact lenses is formed from a crosslinked copolymer of MPC with HEMA and is described in more detail in Section 4.6.2.2. A recent report describes the use of a novel PC-based contact lens formulation in which the cross-linking is achieved by use of a modified PC monomer that is itself a bifunctional cross-linker [Goda & Ishihara 2006; Goda et al. 2006]. The authors claim this allows for significant incorporation of MPC while maintaining mechanical properties.

Bulk hydrogels based upon MPC copolymers have been described that form physical gels by a variety of means, rather than by incorporation of a comonomer to provide chemical cross-linking. Mixture of two MPC copolymers, each containing oppositely charged comonomer components, promotes the formation of polyionic complexes (PICs) of high water contents and excellent biological properties [Ishihara et al. 1994a; Ito et al. 2005; Kimura et al. 2007c]. One particularly interesting system results in spontaneous formation of a hydrogel when MPC copolymers with a comonomer possessing carboxylic acid groups (such as methacrylic acid [MA]) is mixed with MPC copolymer with a hydrophobic comonomer such as BMA [Nam et al. 2002a; Kimura et al. 2004]. This is due to extensive hydrogen bonding by the carboxyl groups, and provided flexibility to control gel properties based on chemical structure, composition, and sequence [Kimura et al. 2005]. Addition of Fe^{3+} to the gel has been shown to enable control of gelation time and viscoelastic properties, without evidence of inflammation in vivo [Kimura et al. 2007a]. These hydrogels have found utility as drug delivery systems [Nam et al. 2002b, 2004b; Kimura et al. 2007b] (see Section 4.6.5). An alternative system uses a water-soluble MPC copolymer containing a vinylphenylboronic acid comonomer; when added to an alcohol-rich material such as polyvinyl alcohol (PVA), a hydrogel is formed which can be dissociated by addition of low molecular weight hydroxyl compounds such as glucose [Konno & Ishihara 2007]. This system is being investigated for the controllable encapsulation of cells such as fibroblasts.

Although these PC methacrylate-based hydrogels are sufficient for devices such as contact lenses where the stresses in use are low, they cannot compete with the mechanical properties possessed by some of the typical biomedical polymers used in the manufacture of medical devices, such as PUs, nylons, and silicone rubbers. Although coating these materials with PC polymers can offer an improvement in their biocompatibility without compromising mechanical performance, there may be instances where a coating is difficult to attain or simply insufficient in durability. One approach to overcoming this issue has been to use polymer blends. A number of studies have reported on the blending of PC polymers, for example, with PUs [Ishihara et al. 1996a,b, 2000; Sawada et al. 2006], in order to generate materials that might be suitable for use in small diameter vascular grafts [Yoneyama et al. 2000]. A more complex approach has been described in which PC end-capped poly(ethylene oxide)-*block*-poly(epsilon caprolactone) has been blended with PU, resulting in migration of the hydrophilic block to the surface interface while the PCL anchors the copolymer firmly in the PU matrix [Zhang et al. 2008a]. Blends of PMPC with polyethylene have been evaluated and suggested for use in place of conventional PVA as a biomaterial [Ishihara et al. 2004]. Similar studies have been described for blends with cellulose acetate for preparing hemofiltration membranes [Ye et al. 2003, 2004] and blends with polysulfone for use in the preparation of hollow fiber membranes for use in the construction of an artifical kidney [Ishihara et al. 1999a,b, 2002; Hasegawa et al. 2001, 2002; Ueda et al. 2006]. PC polymers have blended with poly(lactide-*co*-glycolide) to take advantage of the known biodegradability of the polyester component [Iwasaki et al. 2002]. An interesting feature of many of the blended materials is the tendency for the PC polymer to phase-separate and migrate to an interface within the blend (usually the surface) [Clarke et al. 2001]; it thus becomes predominantly expressed where it is required, at the interface with the body, and less associated within the bulk of the material, hence having little negative impact on the mechanical properties (Figure 4.8). Moreover, the phase separation of the PC polymer is often seen to manifest itself in the formation of domains on the surface, the size and frequency of which can be somewhat controlled to modulate the surface interaction with proteins and cells [Long et al. 2003].

4.5 BIOLOGICAL COMPATIBILITY

4.5.1 Interactions with Proteins, Eukaryotic Cells, and Bacteria

In Section 4.4.1 the antifouling mechanism of action of PC polymer systems was discussed. These properties give rise to the fundamental resistance these materials show

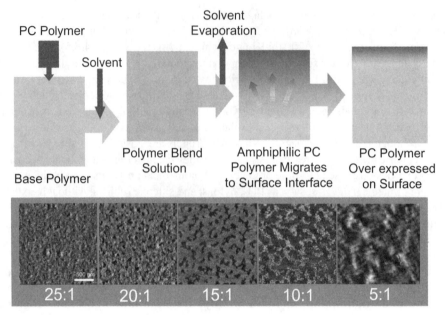

Figure 4.8. Mechanism for surface enrichment of MPC polymers after blending of polymers.

against the irreversible adsorption of many types of proteins, particularly those abundant in plasma and other tissues. Protein adhesion to a surface is required as a first step in the cellular recognition processes that promote cell attachment and proliferation on a substrate. PC polymers therefore possess the ability to suppress cellular adhesion (of particular interest in biomedical application being cellular components of the blood such as platelets, inflammatory cells, and bacteria). The inherent biological compatibility of the materials has therefore been the primary focus for a large number of studies conducted over the past 20 years. Adhesion of proteins such as fibrinogen, human and bovine serum albumin (BSA), γ-globulin and lysozyme, for instance, has been studied by a variety of techniques ranging from standard protein determination assays [Iwasaki et al. 1996b], enzyme immunoassays [Campbell et al. 1994], radioimmunoassay and immunogold labeling [Ishihara et al. 1991], UV spectroscopy [Ishihara et al. 1992], ellipsometry [Murphy et al. 1999, 2000], neutron reflection [Murphy et al. 2000], and spectrofluorimetry [Young et al. 1997a] to name a few. Similarly, the reduction in adhesion of a wide range of cell types has also been reported, including among others, platelets [Campbell et al. 1994; Iwasaki et al. 1994, 1996a], human monocytes [DeFife et al. 1995], macrophages and granulocytes [Goreish et al. 2004], neural cells [Ruiz et al. 1998], fibroblasts [Ishihara et al. 1999c], corneal epithelia [Hsiue et al. 1998], and various pathogenic microorganisms [West et al. 2004; Hirota et al. 2005; Rose et al. 2005; Fujii et al. 2008]. Indeed, protein and cell interaction studies in this area are too numerous to cover here, and there are a number of review articles for an overview of the subject [Lewis 2000, 2004; Nakabayashi & Williams 2003; Nakabayashi & Iwasaki 2004; Iwasaki & Ishihara 2005].

4.5.2 Interaction with Other Tissues

Hemocompatibility is one of the prime considerations for medical devices that contact with blood; resistance to the formation of thrombus is therefore a desirable property for such applications. As we have seen in Sections 4.3.1 and 4.5.1, the biomimetic properties of PC polymers impart the ability to temper the processes leading to thrombus formation, which has lead to their widespread evaluation in cardiovascular applications. Again, there have been a tremendous number of literature reports on the interaction of PC polymers with whole blood [Bird et al. 1989; Ishihara et al. 1992, 1994b, 1996b]. Some workers have attempted to simulate interactions in vivo by use of dynamic flow conditions [Patel et al. 2005], whereas other studies have used ex vivo models in which flowing blood was passed through various PC-coated tubing, with significant reduction in thrombus and platelet adhesion [Chronos et al. 1994]. Ultimately, the exceptional hemocompatibility of PC polymers has found use in a variety of blood-contacting applications (see Section 4.6.1).

In addition to the numerous studies on proteins and blood components, PC polymers have been shown to induce a very low inflammatory response and minimal fibrous capsule formation both in vitro and in vivo in a rabbit intramuscular model [Goreish et al. 2004]. This is supported on the biochemical level by recent studies on PC polymers blended with poly (lactide-*co*-glycolide), which have shown the presence of PC reduces the expression of mRNA of the inflammatory cytokine interleukin (IL)-1-beta in adherent human premyelocytic leukemia cells [Iwasaki et al. 2002]. Some of the most extensive data on tissue compatibility has been generated on the PC-coated coronary stents in various coronary artery models. A number of studies have examined the tissue reaction in the arterial wall associated with the coating and have all concluded that the extent of tissue reaction is minimal, the coating being very well tolerated [Kuiper et al. 1998; Whelan et al. 2000; Malik et al. 2001]. Moreover, in a study designed to quantify the extent of inflammation generated from stent implantation comparing PC-coated stents with PC-coated stent containing anti-inflammatory drugs, it was difficult to detect a meaningful difference between the two, given that the PC coating alone invoked such a low inflammatory response in the vessel [Huang et al. 2003].

4.6. APPLICATIONS OF PC POLYMERS

4.6.1 Cardiovascular Applications

One of the first commercial applications of a PC polymer coating in the cardiovascular area was the use of a simple MPC-*co*-LMA copolymer on coronary guide wires. In a clinical assessment, 5 different coated wires were used in 50 consecutive angioplasty procedures [Gobeil et al. 2002]. Significant thrombus formation was a frequent finding on the angioplasty guide wires, occurring in 48% of cases; only the PC polymer-coated guide wires showed no thrombus formation at all. With this sort of striking result, it is not surprising that there has been further evaluation of these materials on other cardio-vascular devices, including ventricular assist devices [Kobayashi et al. 2005; Snyder et al. 2007], extracorporeal circuits, vascular grafts and, most notably, coronary stents, some of which are now discussed in more detail.

4.6.1.1 PC-Coated Coronary Stents.

The crosslinked PC polymer described in Section 4.4.2 was first used as an ultrathin coating some 100 nm thick on the Bio*divYsio*™ range of coronary stents and subsequently Conformité Européene (CE) mark approved and U.S. Food and Drug Administration (FDA) cleared. A number of open registries evaluated the early clinical performance of the PC-coated Bio*divYsio* stent and the first formal trial was SOPHOS, powered to demonstrate the product's safety and efficacy [Boland et al. 2000]. The coating was also applied to stents specifically designed for use in small vessels of reference diameter 2.0–2.74 mm and shown to be safe and effective [Bakhai et al. 2005]. FDA approval came following the DISTINCT trial; the first randomized multicentre clinical trial comparing a PC-coated Bio*divYsio* stent (313pts) with an uncoated stent (MULTI-LINK DUET®, 309pts, Guidant Corporation, Indianapolis, IN). Interestingly, in this study there was no subacute thrombosis (SAT) reported for the PC-coated stent. A complete overview of the in vitro, preclinical and clinical studies performed on the PC-coated stent can be found elsewhere [Lewis & Stratford 2002].

Two versions of the Bio*divYsio* stent possessing thicker PC-based polymer coatings (~1 μm) were developed to enable drug uptake and delivery [Lewis & Vick 2001]: the Matrix LO™ coronary stent, designed for the loading and release of small therapeutic molecules <1200 Da (Figure 4.9), and the Matrix HI™ coronary stent, coated with a cationically modified PC polymer to allow interaction and delivery with negatively charged biomacromolecules such as heparin and nucleic acids. A wide range of studies on these stent platforms in combination with different potential antirestenotic therapeutics has therefore been possible, some of which are reported in Section 4.6.7.1; this section will focus on those studies that have lead to significant clinical study and commercialization of a product. Huang and coworkers reported on the reduction in inflammation and neointimal hyperplasia in a porcine coronary artery model using a PC-coated Bio*divYsio* stent that delivered a steroidal anti-inflammatory agent [Huang et al. 2003]. These promising preclinical results lead to the commercialization of the Dexamet™ stent in 2001; this product was composed of the crosslinked PC coating, preloaded with the anti-inflammatory drug dexamethasone. This was the second drug eluting stent to reach the market and has continued to report good results in clinical studies [Shin et al. 2005; Han et al. 2006]. The polymer has subsequently been used by Abbott Vascular Devices, in the development of the Zomaxx™ stent [Burke et al. 2006; Abizaid et al. 2007]. In addition, it is used by Medtronic in the Endeavor® stent and this has involved further extensive preclinical and clinical evaluation [Fajadet et al. 2006, 2007; Kandzari & Leon 2006; Kandzari et al. 2006; Gershlick et al. 2007].

4.6.1.2 Vascular Grafts.

A vascular graft may be biological or synthetic and is a conduit to restore blood flow by bypassing a diseased vessel. They are commonly made of expanded poly(tetrafluoroethylene) (ePTFE) or Dacron (poly(ethylene tere-

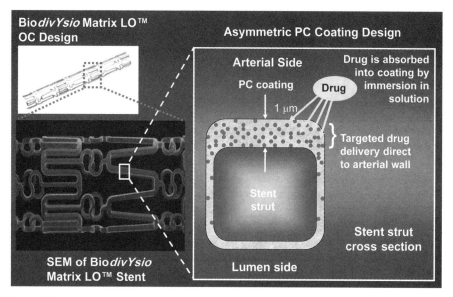

Figure 4.9. MPC polymer-coated Bio*divYsio* coronary stent (top left), section under scanning electron microscopy (SEM) (bottom left) and schematic cross section across the stent strut for the Matrix LO drug delivery coating.

phthalate), PET) and are are used to create a bypass around diseased arteries or replace damaged vessels. There are currently no suitable materials for use in the manufacture of narrow bore vascular grafts (less than 3 mm), as conventional materials thrombose and block. The most common complications associated with their use are occlusion, infection, hyperplasia at the anastomoses, distal embolization, and erosion of adjacent structures. It has been proposed that there could be potential benefit by application of PC polymers to synthetic grafts. Use of PC polymer coatings in combination with such devices is shown to be of benefit, not only from the nonthrombogenicity bestowed upon the device [Chevallier et al. 2005], but also from the fact that studies in a canine model showed reduced intimal hyperplasia at the anastamoses where synthetic graft is sutured to the tissue [Chen et al. 1997, 1998]. Others have blended base PU with PC copolymer variants (usually copolymerized with a monomer to aid in interaction of the two blend components) in order to improve biological performance without unduly affecting the graft's physical characteristics [Yoneyama et al. 2000] (see Section 4.4.2).

4.6.1.3   *Extracorporeal Circuits.* An extracorporeal procedure is one in which blood is taken from a patient's circulation in order to perform some process before returning it to the patient. The apparatus that carries the blood outside the body is known as the extracorporeal circuit and can include the various tubing and connections, through to a specific device for instance to achieve: hemodialysis, hemofiltration, plasmapheresis, apheresis, or cardiopulmonary bypass, including extracorporeal membrane oxygenation (ECMO). As the components of the circuits contact the blood, it is not surprising that manufacturers have sought materials and surface treatments that improve hemocompatibility. The simple physisorbed MPC-*co*-LMA copolymer system described in Section 4.4 has found great utility in this application; its flexibility to be applied as a stable coating to the wide range of materials that are used in circuit construction has made it an appealing choice for commercialization. Sorin Biomedica have applied the technology, termed "Mimesys™," to a range of its extracorporeal devices and have demonstrated advantages in platelet preservation [Myers et al. 2003]. It has also been shown in a sizable clinical study in moderate risk patients undergoing coronary revascularization with cardiopulmonary bypass that systemic heparin dose can be reduced by use of a PC-coated circuit and results in significant improvement in patient outcomes [Ranucci et al. 2004]. In a further study of two groups of 10 patients undergoing elective cardiac operations for different congential anomalies, the whole extracorporeal circuit, including cannulas and tubing, was PC coated for one of the groups [De Somer et al. 2000]. The PC-coated circuit had a favorable effect on blood platelets when studying the changes during cardiopulmonary bypass. A steady increase of thromboxane B2 and β-thromboglobulin (markers of platelet activation) was observed in the control group, whereas plateau formation was observed in the PC-coated group. Complement activation was also lower in the PC-coated group. The authors concluded that clinically, this effect may contribute to reduced blood loss and less thromboembolic complications. Similarly, Pappalardo et al. [2006] recently reported on the effects of PC-coated oxygenators on blood coagulation in patients; they found that for patients undergoing high-risk open heart surgery and receiving tranexamic acid (to reduce fibrinolysis and blood loss), a PC-coated oxygenator may reduce intraoperative thrombin formation and the associated consumption of platelets, fibrinogen, and antithrombin. These examples illustrate the mounting positive clinical data that support the use of PC-coated circuits for use in blood-contacting applications.

4.6.2   Ophthalmic Applications

4.6.2.1   Intraocular Lenses.   Intraocular lenses (IOLs) are implants for the eye that replace the crystalline lens as refractive surgery to correct the optical power of the eye or restore vision as the original lens has been obscured by a cataract. Although originally based on poly(methylmethacrylate) (PMMA) new IOL designs are made of foldable acrylic and silicone-based materials. While the physical properties are enhanced, these materials can still induce an inflammatory response and suffer from posterior capsular opacification due to their incompatibility with the ocular environment. This drawback prompted early investigations into the use of PC-based materials to improve the biological properties of IOLs, both as surface treatments and as bulk materials for the lens fabrication. Significant reductions in the adhesion of proteins such as fibrinogen, various mammalian cells such as lens epithelia and fibroblasts, macrophages and granulocytes, and bacteria such as *Staphylococcus epidermidis* have been reported [Lloyd et al. 1999].

More recently, activity has once again increased in the IOL area and MPC-modified acrylic lenses have been shown to reduce foreign body reaction and suppress adhesion and proliferation of lens epithelial cells (LECs) [Okajima et al. 2005, 2006b]. Further studies have demonstrated that both fibroblasts and bacterial adhesion can be reduced, potentially reducing the risk of endophthalimitis [Shigeta et al. 2006; Okajima et al. 2006a]. Others have taken silicone IOLs and tethered MPC to the surface using an air plasma treatment; this led to improved hydrophilicity of the lens surface and reduction in platelet, macrophage, and LEC adhesion on the surface [Yao et al. 2006].

4.6.2.2 Contact Lenses. The ocular environment has been another major area of commercial application for PC polymers [Goda & Ishihara 2006]. Contact lenses lay on the surface of the cornea where they act as a barrier to oxygen diffusion and present a persistent mechanical irritation. The Proclear family of soft contact lenses is based on omafilcon A, an MPC-based polymer that provides a hydrated gel matrix into which oxygen can dissolve and diffuse to the cornea surface. The proteins and lipids that comprise the tear fluid can transport through the matrix without fouling the lens, providing nutrients and protection against bacterial infection [Young et al. 1997a,b] and the materials have been shown to reduce adhesion of an ophthalmic clinical isolate of *Pseudomonas aeruginosa*, macrophages, granulocytes, and corneal epithelial cells [Lloyd et al. 2000; Andrews et al. 2001]. The highly hydrophilic nature of the MPC holds the water in the lens and on its surface, allowing the tear film to wet the lens and not break up, causing islands for potential deposition of fouling species (see Figure 4.10). This leads to reduced on-eye dehydration of the lens, which is particularly useful for dry eye sufferers. These types of lenses can be fabricated by direct cast moulding in which the monomer formulation is polymerized directly within a mould that creates the desired shape of the lens.

Extended wear lenses require oxygen permeability that is beyond the capabilities of conventional hydrogel materials. Although MPC-based lenses possess some of the highest water contents of conventional soft lenses, they are not recommended for continual wear over a month. For this property, silicone hydrogel materials are used for their excellent oxygen transmissibility. Where a surface treatment is necessary to improve biocompatibility, an in-mould coating (IMC) technique has been described where a suitably functionalized MPC polymer is applied to the inside surfaces of a

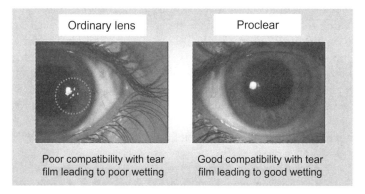

Ordinary lens — Proclear

Poor compatibility with tear film leading to poor wetting — Good compatibility with tear film leading to good wetting

Figure 4.10. Poor tear wetting on a conventional soft lens surface compared to an MPC polymer contact lens (Proclear).

mould, whereupon the mould is filled with an appropriate polymer formulation and cured [Court et al. 2001]. As the functional groups within the mould react, the reactive functionalities of the MPC polymer become grafted into the surface of the forming article, producing a highly stable, wettable, and biocompatible layer.

4.6.2.3 Other Ocular Devices. Glaucoma drainage devices (GDDs) are used to reduce the pressure in the eye that results as a consequence of glaucoma. GDDs are available that act as a conduit to relieve the pressure but are prone to blockage. GDDs that have been coated with MPC polymers have been shown to not only cause less damage to corneal endothelial cells [Lim 2003] but also to reduce adhesion of proteins, human fibroblasts, and scleral macrophages, suggesting a benefit in terms of drain failure [Lim 1999]. The puncta are the openings of the tear ducts on the eyelid margin and punctal plugs are devices used to prevent tears draining into the nose *via* the tear ducts; stemming the tear flow is a strategy to maintain the tears in the eye for a longer time to overcome dry eye syndrome. Grace Medical is developing a range of PC-coated silicone punctual plugs to provide a hydrated layer for comfort and improved biological interaction, with lower propensity for infection.

4.6.3 Anti-Infective Applications

As discussed in Section 4.5.1, there is growing evidence to support the use of PC coatings in devices where bacterial colonization is the primary mode of failure. In addition to the ocular devices already discussed in Section 4.6.2, there are a number of other commercial devices that take advantage of this property in order to enhance their clinical performance.

4.6.3.1 Urological Devices. The field of urology is another area in which PC polymers may find utility given their inherent ability to prevent or reduce the extent of adhesion of certain bacteria. Bacteria must first adhere in order to reproduce; adhesion can be mediated by protein and polysaccharide excretion. There are numerous reports in the literature on the ability of PC-based polymer systems to resist the adhesion of various strains of bacteria (see Lewis 2000 for an overview). Urological devices are often indwelling for long periods and are prone to infection from such organisms as *Escherichia coli*; *Proteus mirabilis* colonization results in biofilm formation that promotes mineralization by struvite and hydroxyapatite, leading to encrustation and occlusion of the device lumen. Laboratory tests that mimic the encrustation process have shown large reductions in mineralization for PC-coated latex and silicone substrates. This is further supported by a limited clinical study which showed reduction in biofilm formation on a PC-coated PU ureteric stent from 70% to 30% compared to uncoated control 3 months postimplantation [Russell 2000].

4.6.3.2 Tympanostomy Tubes. The inherent nonfouling properties of PC materials are useful in maintaining small lumen devices clear of biological occlusion or infection. Tympanostomy tubes are small conduits that are inserted through the eardrum to treat persistent middle ear effusion or serous otitis media in children. A recent commercial development has been the PC-coated tympanostomy tubes, which are either fluoroplastic (Gyrus ENT) or silicone (Grace Medical) devices coated with the crosslinkable PC polymer described in Section 4.4.1. The coating has demonstrated the ability to resist staphylococcal and pseudomonal biofilm formation equally as well as antibacterial silver-based systems [Berry et al. 2000].

4.6.4   Orthopedic Applications

Although the rationale for their use may not appear immediately obvious, PC polymers are also finding utility in the field of orthopedics. MPC has been successfully grafted onto the surface of crosslinked polyethylene to produce a highly lubricious bearing layer for enhanced wear resistance of artificial hip joints. At high density, the MPC layer displayed a marked decrease in friction and produced significantly less wear particles and wear of the liner than the untreated surface after 10^7 cycles in a hip joint simulator test [Moro et al. 2004, 2006]. Further studies suggest high graft density or further crosslinking by gamma sterilization is important for maintaining the longevity of the surface [Kyomoto et al. 2007b, 2008]. A similar grafting technique has also been applied to cobalt-chromium-molybdenum (Co-Cr-Mo) alloy, again resulting in a marked reduction of the coefficient of friction for the grafted surface [Kyomoto et al. 2007a]. In contrast to the use of PC polymers as lubricity enhancers, others have investigated modified coatings for their bioactive potential to induce bone formation. Initial work by Habib and coworkers using a cationically modified PC polymer had suggested that the coating may encourage human osteoblasts to adhere and mineralize [Habib et al. 2003]. This, however, was not borne out in vivo in a rat tibia model, where PC-coated pins produced similar amounts of bone and marrow apposition compared to uncoated controls.

4.6.5   Biosensors and Diagnostics

In-dwelling biosensor systems suffer from long-term loss of sensitivity due to fouling of the sensing window surface. Any modification to prevent this occurrence must not interfere with the sensor function (i.e., must remain transparent to the appropriate wavelength of light for an optosensor and must maintain sensor response time). PC polymers present an obvious opportunity for the improvement of implantable sensor biocompatibility due to their proven resistance to protein and cell adhesion [Abraham et al. 2004, 2005]. MPC-*co*-BMA polymers have been used as a sensor membrane to improve the hemocompatibility of an intravascular pO_2 sensor, demonstrating stable behavior in static blood compared to PU and PVC membranes which deteriorated over time [Zhang et al. 1996]. A similar copolymer was modified with components that enhance affinity to PU in order to improve the long-term attachment for use with an electrochemical enzyme-based glucose sensor [Yang et al. 2000]. Similarly, antibiofouling MPC-*co*-LMA copolymers have been used to coat luminescent oxygen sensors [Navarro et al. 2001]. Nanometer-sized coatings of such materials were shown to reduce significantly the adhesion of marine bacteria (more than 70%) and thrombocytes (more than 90%) to the surface of tris-(4,7-diphenyl-1,10-phenanthroline)ruthenium(II)-doped silicone layers. These novel biofouling-resistant optosensors were successfully validated against a commercial oxygen electrode and were shown to respond faster than the electrochemical device for large oxygen concentration changes. The biomimetic coatings were thought to be particularly useful for drift-free long-term operation of environmental optosensors and in vivo fiber-optic oxygen analyzers. Similar polymer systems have also shown good utility in the preparation of miniaturized glucose sensors [Chen et al. 1992, 1999; Kudo et al. 2008], including a novel flexible PDMS-based system coated with MPC-*co*-LMA and fabricated using microelectromechanical systems (MEMSs) techniques [Kudo et al. 2006]. A copolymer of MPC with glycidyl methacrylate has been used to modify epoxy coatings for optical biosensors [Chae et al. 2007]. These materials were optically transparent, had excellent adhesive properties, and were highly antifouling against microorganisms.

There has been a great deal of interest in the use of PC polymers in diagnostic uses because of the ability of the polymer to resist nonspecific protein interactions. Copolymers of MPC with dimethylaminoethyl methacrylate (DMA) have been used in combination with HAuCl$_4$ to protonate the DMA and the remaining tertiary amine reduces the AuCl$_4$ counterion to zerovalent gold in situ. This results in MPC-stabilized gold nanoparticles with potentially improved biocompatibility in diagnostic applications [Yuan et al. 2006b; Jin et al. 2008]. MPC copolymers with glycerol methacrylate have been shown to stabilize magnetite sols by chemical coprecipitation of ferric and ferrous salts in the presence of the polymer, forming nanoparticulates of potential utility in magnetic resonance imaging [Yuan et al. 2006a]. Copolymers possessing reactive ester side groups have been reacted with proteins such as trypsin and immobilized on PMMA microchips to allow rapid enzyme catalyzed hydrolysis and separation of product from substrate [Sakai-Kato et al. 2004]. PC polymers possessing *p*-nitrophenyl ester groups have been used as a surface modifier on nanoparticles onto which luciferase has been subsequently immobilized. The nanoparticles were able to react with ATP, luciferin, and oxygen and showed promise as a photochemical-sensing microdiagnostic system [Konno et al. 2006]. An alternative approach has been to make PC-bearing nanoparticles of poly(L-lactic acid) (PLA) for the immobilization of antibodies and enzymes [Ito et al. 2006]. It was demonstrated that the nanoparticles aggregated upon addition of antigen or could catalyze enzyme reactions with their substrate when in suspension; it was concluded these systems could be of use in nano-/micro-scaled diagnostic systems. In yet another embodiment of this concept, MPC nanoparticles coated with anti-bisphenol A antibodies were applied in a monodisperse layer onto a piezoelectric immunosensor, which demonstrated significantly improved sensitivity attributed to the colloidal stability of the nanoparticles [Park et al. 2006]. Similar PC polymer systems have been used to immobilize antibodies to the gold electrode of a quartz crystal microbalance (QCM) through thiol linkages [Park et al. 2007]; these have demonstrated excellent resistance to nonspecific binding and a reduction in the immunologic activity of the antibody. Copolymers of MPC-BMA and *p*-nitrophenyloxycarbonyl poly(ethyleneglycol) methacrylate have been used to conjugate antibodies to enhance specific signals for selective and reliable immunoassays [Nishizawa et al. 2008]. In an effort to control the surface properties further, MPC copolymer brushes have been formed by surface initiated ATRP and antibodies subsequently immobilized for use in highly sensitive enzyme-linked immunosorbent assays (ELISA) [Poilane et al. 2007].

Recently, there have been a number of reports on the use of PC polymers for use in the construction of microfluidic chips. In a simple one-step approach, various MPC copolymers could be coated onto PMMA [Bi et al. 2006] or PDMS [Sibarani et al. 2007a] microfluidic chips. The coated chips exhibited more stable electroosmotic mobility, reduced nonspecific adhesion of serum proteins and plasma platelets, and were useful for the electrophoresis of proteins. One such PC copolymer variant was made net-negatively charged by incorporation of potassium 3-methacryloyloxypropyl sulfonate, and 3-methacryloyloxypropyl trimethoxysilane was added to enable covalent attachment to the substrate (see Section 4.4.1). This was subsequently used to coat silica-based microchannels to control the electroosmotic flow while suppressing nonspecific protein adsorption *via* a one-step surface modification [Xu et al. 2007a,b]. In addition to physical coatings, MPC brushes have been applied to microfluidic device channels by surface-initiated graft polymerization with much the same properties as the physical coatings regarding protein adsorption [Sibarani et al. 2007b].

4.6.6 Separation Systems

Membranes for separation processes are required to reject species of a particular size (dependent upon the choice of membrane) while maintaining a practical flux across the membrane. The most common issue for the use of membrane technologies is the fouling of the membrane surface and subsequent decline in flux. It was recognized quite early on that there could be value in the development of PC-based treatments for biofiltration membranes which could prevent or inhibit protein fouling. Microfiltration membranes of cellulose triacetate, polyether sulfone, and polyvinylidene fluoride were etched with oxygen in a plasma chamber to generate surface hydroxyl groups and were then treated with the MPC [Dudley et al. 1993; Reuben et al. 1995]. These membranes were evaluated with water, buffer, BSA, yeast fermentation broth, beer, and orange juice, and were seen to increase the flux and decrease the rate of fouling in most cases. Recently, an interesting application was reported for the use of MPC polymers in chromatographic separations. Jiang and coworkers have described the use of porous silica particles that have been graft polymerized with MPC to produce a novel zwitterionic stationary phase for the separation of peptides by hydrophilic interaction liquid chromatography (HILIC) [Jiang et al. 2006].

More relevant to biomedical applications is the filtration of blood. MPC copolymers have been used to coat PET blood filtration mats with an intention to reduce the amount of platelet adhesion and activation. One immediate problem associated with the polymers was the wettability; the blood is required to permeate through the filter without the need for physical priming. Coating compositions containing water were seen to overcome the problem but did not offer a practical commercial solution [Lewis et al. 2000] (see Section 4.3.2). MPC copolymers that were crosslinkable and contained low amounts of hydrophobic component were seen to offer a spontaneously wettable surface treatment [Lewis et al. 2003]. Another approach focused on the surface mobility of the MPC group as a key factor for blood compatibility and aimed to increase the distance of the MPC group from the backbone using a diethylene oxide bridge, in order to improve platelet compatibility on PET nonwoven filtration fabrics [Iwasaki et al.

2003]. It is therefore clear that the performance of hemopurification membranes can be significantly improved by use of strategies to reduce protein adhesion and clot formation. In a more commercially acceptable approach, cellulose acetate hollow fiber membranes have been modified with MPC-*co*-BMA by blending the polymers (see Section 4.4.2) in a dry-jet wet spinning process [Ye et al. 2003, 2004, 2005]. The resulting blends showed good permselectivity, low protein adhesion, platelet adhesion, and fouling properties during permeability experiments.

A number of strategies have been described in which PC polymers have been used in the separation of cells. The simplest has been to mix a range of water-soluble polymers with whole blood in order to aggregate and separate the platelets as an alternative to centrifugation [Sumida et al. 2006]. A number of polymers, including the MPC-*co*-BMA polymer were effective at separating the platelets; only the PC polymer, however, did not activate the platelets. Another approach has been to coat microcanals with a PC PIC. By changing the surface electrical potential, cells were separated from whole blood by attraction to the net positively charged surface, and the cell morphology maintained by the presence of the PC [Ito et al. 2005]. The charged nature of these PC-based PICs has also presented the opportunity to investigate the materials for tissue adhesion properties using a dura incision model, with the adhesive properties being dependent upon the polymer concentration and water content of the hydrogel [Kimura et al. 2007c].

A novel PLA nanoparticle system modified with an MPC-BMA copolymer with *p*-nitrophenyloxycarbonyl poly(ethylene glycol) methacrylate for immobilization of antibodies has been reported. The antibody allows for highly selective binding to target species for specific protein separation applications [Goto et al. 2008]. Well-defined polymer brushes of MPC and glycidyl methacrylate prepared by surface-initiated ATRP have been subsequently reacted with pyridyl disulfide to enable immobilization of antibody fragments on the brush surface [Iwata et al. 2008]. Antigen/antibody binding was shown to be greater than for non-MPC control surfaces, offering good potential for specific separations.

4.6.7   PC Polymers for Drug Delivery

It is interesting that although phospholipids have been used for many years in the application of drug delivery (e.g., liposomal systems), it is only relatively recently that this has become a major focus for the use of PC-based polymers [Lewis 2006]. The main reason for this is that some of the more complex polymer architectures that enable formation of self-assembled structures or stimuli-responsive properties have only been possible by use of more controlled polymerization methods such as ATRP and RAFT (see Section 4.2). Broadly, the areas of utility can be divided into coating-modulated delivery, delivery from gels systems, nano/microparticulates, and drug conjugates.

4.6.7.1   Drug Delivery Coatings. Biocompatible surface coatings that can hold and modulate release of active agents has become an area of increasing research interest; fueled by the development of drug-device combination products, most notably illustrated by the drug eluting stents (Section 4.6.1.1). Although McNair first described the use of PC hydrogel coatings for drug delivery purposes [McNair 1996], it was some

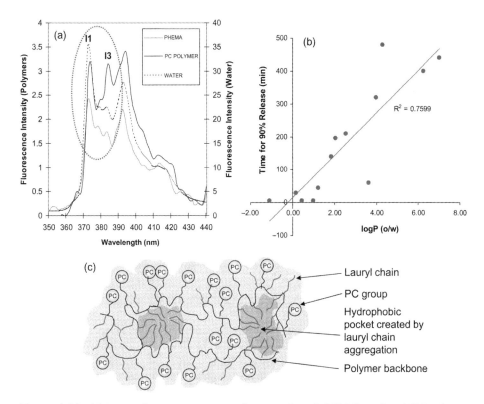

Figure 4.11. (a) Pyrene fluorescence spectra in water, in poly(HEMA), and an MPC polymer coating; the ratio of I1:I3 demonstrates the presence of a hydrophobic environment for the MPC polymer. (b) Relationship between the log P (o/w) of a range of therapeutic agents and the time for 90% release from a crosslinked MPC coating. (c) Schematic representation of the hydrophobic pockets formed within the polymer structure.

years later before a full evaluation of the crosslinked PC polymer system described in Section 4.1 was more fully described [Lewis et al. 2001b]. Membranes of the polymer were fabricated and characterized in terms of water content and molecular weight cut-off, indicating that species greater than 1200 Da could not diffuse into the polymer network when fully cross-linked. The hydrophobic probe pyrene was used to demonstrate the presence of hydrophobic domains formed from the aggregation of the LMA component (Figure 4.11a), which can act as reservoirs for interaction with hydrophobic drugs (Figure 4.11c). Evaluation of the in vitro loading and elution of a wide range of different therapeutics showed the drug's rate of release was a function of its water solubility and also its oil/water partition coefficient (Figure 4.11b). This finding was explained in terms of the more hydrophobic drugs partitioning into, and interacting with, the hydrophobic domains of the polymer coating. Others have emulated this system, replacing LMA with the more hydrophobic and crystalline monomer, stearyl methacrylate [Fan et al. 2007]. Evaluation of coatings of this material on a stent confirmed that it retained its biocompatibility and could release rapamycin in a controlled manner. These PC polymer systems therefore offer a very flexible approach to drug delivery where crosslinked density and degree of hydrophobicity can be tuned to control the release characteristics of a particular therapeutic agent.

A number of studies have appeared in the literature whereby the PC-coated stent described in Section 6.1 has been used as a platform to evaluate the delivery and efficacy of a variety of potential antirestenotic therapies. Irinotecan eluting PC-coated stents were evaluated in the aortas of hypercholesterolemic rabbits and at high dose were found to inhibit neointimal hyperplasia while decreasing inflammatory infiltrate and media necrosis [Berrocal et al. 2006]. New and coworkers have described the use of the PC-coated stent to deliver 17-β estradiol [New et al. 2002]; they observed significant reduction in neointimal formation without effect on the endothelial regeneration in the pig model of in-stent restenosis, possibly by inhibition of extracellular signal-regulated kinase (ERK) activation in smooth muscle cells [Han et al. 2007]. This has not, however, translated into a clinical benefit, as a recently reported randomized study showed no superiority over the PC-coated control [Airoldi et al. 2005]. Angiopeptin, an inhibitor of smooth muscle cell proliferation, was evaluated ex vivo in human saphenous vein segments and in vivo in pig coronary arteries [Armstrong et al. 2002]. ^{125}I-labeled angiopeptin was used to visualize local delivery into the arterial sections, and although no adverse tissue reaction was seen from any of the PC-coated stents, no significant differences in neointimal formation or luminal cross section were observed. This drug–device combination was evaluated in a small clinical study with 14 patients and appeared feasible and safe with a modest degree of neointimal hyperplasia [Kwok et al. 2005]. Some of the most impressive results from a PC-coated stent/drug combination have been obtained using Zotarolimus (ABT-578), a rapamycin analogue with immunosuppressive and antiproliferative activity. Collingwood et al. described their work in the porcine overstretch model and noted significant inhibition of neointimal formation with no giant cell reaction at 28 days [Collingwood et al. 2005]. These few examples aid to illustrate the broad use of the crosslinkable PC polymer coating as a flexible drug delivery matrix for use in biomedical applications.

A cationically modified crosslinked PC polymer was developed [Lewis et al. 2004a], primarily for use as a coating on a coronary stent, for the delivery of large biomacromolecular species such as DNA (see Section 6.1) [Palmer et al. 2004]. While the cationic functionality provides a vehicle to permit ion exchange capability, a balance must be struck between this and adverse effects on the biological properties of the coating [Rose et al. 2004, 2005]. This platform has been evaluated for the delivery of c-myc antisense oligodeoxynucleotide (AS-ODN) and demonstrated minimal systemic delivery of the therapeutic and an associated reduction of in-stent neointimal hyperplasia in a porcine coronary artery model [Chan et al. 2007]. In a modification of the coating procedure, a layer-by-layer approach has been described of alternating cationic MPC polymer and c-myc AS-ODN, which affords a more controlled and sustained release profile of AS-ODN elution [Zhang et al. 2008b]. Others have delivered oligonucleotides (ODNs) containing the radionucleotide ^{31}P for the local radiotherapy in the arterial wall and have demonstrated extended delivery to the tissue over a 2-week period [Lewis & Vick 2001]. Similar studies have been carried out using small decoy ODNs, but although successful stent loading and ex vivo deposition and nuclear uptake in the vessel wall were reported, in vivo vascular ODN transfer was not achieved [Radke et al. 2005]. More encouraging were results from the stent-based delivery of naked plasmid DNA encoding for human vascular endothelial factor-2 (VEGF-2) [Walter et al. 2004]. In this study performed in hypercholesterolemic rabbits, re-endothelialization of the stented vessel wall was accelerated via the VEGF-2 gene eluting stent. It can be concluded that the cationically modified PC coating is a useful platform from which to deliver a range of large biomacromolecules and has found particular utility in delivery of nucleic acid species.

4.6.7.2 Gel-Based Drug Delivery Systems. A family of spontaneously forming hydrogels that gel upon combination of a carboxyl-rich MPC copolymer with MPC-*co*-BMA has been discussed briefly in Section 4.4.2. These systems have proven useful for the pH and erosion-dependent delivery of peptides and proteins such as insulin and cytochrome c [Nam et al. 2002b, 2004a,b]. Related systems can be tailored by balancing hydrophilicity and hydrophobicity with inclusion of ionic groups and aromatic components to provide reservoirs for interaction of a number of useful therapeutic molecules including 5-fluorouracil, indomethacin, ketoprofen, and doxorubicin [Kimura et al. 2007b].

ATRP has been used to fabricate MPC-based ABC and ABA triblock copolymer gelators with a variety of comonomers to confer the ability to initiate gelation upon change in pH or temperature [Castelletto et al. 2004; C. Li et al. 2005a,b; Y. Li et al. 2005] (Figure 4.12). If an initiator containing a disulfide linkage is used in the synthesis, an ABA triblock can be formed that has a biochemically degradable bond on the middle of the B (MPC) block [Li et al. 2006]. Upon gelation, a micellar gel is formed that contains hydrophobic domains bridged with MPC chains; upon biochemical cleavage of the disulfide, the MPC linkage is cleaved, and the micellar gel is broken, releasing thiol-terminated MPC micelles. This system has shown great promise as a wound dressing in a 3D tissue engineered skin model [Madsen et al. 2008].

4.6.7.3 Nano/Micro Particulate Drug and Gene Delivery. Probably the simplest demonstration of the utility a PC polymer for drug delivery was the reported use of a water-dispersible MPC-*co*-BMA copolymer containing 70% BMA that can be used to solubilize paclitaxel (PTX) [Konno et al. 2003]. The resulting polymer–drug aggregates are some 50 nm in diameter and the PTX solubility is increased from 0.1 to 5 mg/mL. This formulation has been shown to have significant antitumoral activity in vitro and to be well tolerated in vivo [Wada et al. 2007]. A slightly more sophisticated approach was taken by Yusa et al., whereby the homopolymer of MPC was synthesized in water by RAFT-controlled radical polymerization and end-capped with a dithioester moiety. Using the dithioester-capped PMPC as a macro chain transfer agent, AB diblock copolymers of MPC and BMA were synthesized which self-assemble to form micelles that can solubilize PTX efficiently [Yusa et al. 2005]. Water-soluble MPC polymers bearing hydrazide groups have been prepared by conventional radical polymerization and used as an emulsifier to form nanoparticles of ~200 nm in diameter. The hydrazide has been shown to react with unnatural carbohydrate groups generated on

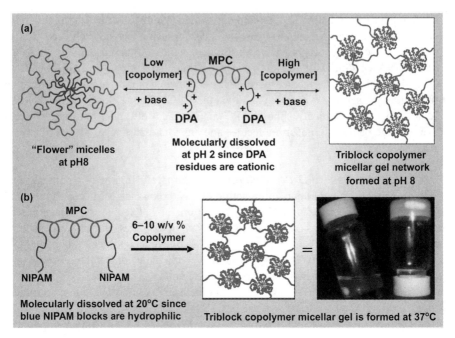

Figure 4.12. Mechanisms of gelation for (a) a pH-responsive MPC ABA triblock copolymer; (b) a temperature-responsive MPC ABA triblock copolymer with NIPAM, N-isopropylacrylamide. Courtesy of Prof. Steve Armes, University of Sheffield.

human cervical cancer cell surfaces, acting as a targeting method to deliver doxorubicin or PTX-loaded nanoparticles into the cells [Iwasaki et al. 2007a].

ATRP has been used to great effect in the construction of self-assembling structures based upon MPC that have potential as drug delivery systems. A cholesterol-based macroinitiator has been used to make an MPC block copolymer with surfactant properties [Xu et al. 2005b]. It has been shown that, at the critical micelle concentration (cmc) in water, micelles form that can be loaded with anticancer agents such as doxorubicin which is released over several days [Xu et al. 2005a]. In recent work, poly(D,L-lactide) has been modified to produce a bromine macroinitiator and used in the preparation of amphiphilic diblock copolymers with MPC [Hsiue et al. 2007]; this ensures a biodegradable hydrophobic block which has demonstrated good cytocompatibility. Single-walled carbon nanotubes have also been successfully functionalized with MPC using surface initiated ATRP, enabling their aqueous dispersion and potential use as a drug delivery system [Narain et al. 2006].

ATRP has also been used to make pH-responsive AB diblocks of MPC and various amines comonomers such as dimethyl-, diethyl- and diisopropylaminoethyl methacrylates (DMA, DEA, and DPA) [Ma et al. 2003], whereby at neutral pH, the amine block is unprotonated and hydrophobic in nature. This induces the micellization of the polymer [Mu et al. 2006] and provides a hydrophobic core for carrying drugs, such as dipyridamole [Salvage et al. 2005; Giacomelli et al. 2006]. At lower pH (e.g., inside

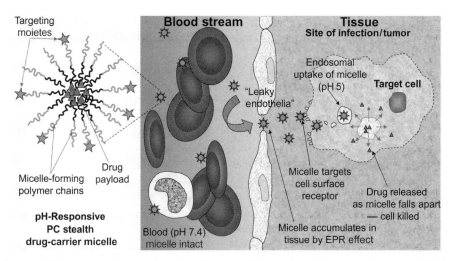

Figure 4.13. Proposed mechanism for cell entry and drug release for a targeting MPC diblock copolymer drug-carrying micelle system. EPR, enhanced permeability and retention.

the endosome once taken into a cell), the amine becomes protonated, and the diblock disperses to become molecularly dissolved, delivering its payload to the cell. Moreover, these AB diblocks can be functionalized with moieties such as folic acid, so that the resulting micelles possess cell-targeting ligands on their surfaces which can aid the specificity of their uptake [Licciardi et al. 2005, 2006, 2008] (Figure 4.13). By controlling the relative lengths of the A and B blocks, these systems can be made to form pH-responsive vesicle structures, or polymersomes, as opposed to micelles, wherein water-soluble drugs such as doxorubicin hydrochloride can be encapsulated into the aqueous core [Du et al. 2005].

The use of cationic components in the MPC copolymers confers upon the polymer the ability to condense nucleic acids such as DNA into polyplexes that could be useful as gene transfection agents. The pH-responsive MPC AB diblocks described earlier can adsorb to surfaces to form a layer that if subsequently contacted with a solution containing DNA, can adsorb it from solution [Zhao et al. 2005]. In solution, the polymers form polyplexes of various shapes and sizes dependent upon block composition [Mu et al. 2004; Chim et al. 2005] and which have been shown to be capable of transfecting cells in vitro and stabilize the DNA from enzymatic degradation [Lam et al. 2004; Zhao et al. 2007, 2008]. In a particularly elegant system, Battaglia et al. has described the use of these MPC AB blocks that form polymersomes to encapsulate DNA. These systems appear to be taken into cells very rapidly and once in the endosomal compartment, the polymersome dissolves as the amine component becomes protonated. As the polymer takes on a net positive charge, it interacts with the enclosed DNA to form a polyplex, protecting the DNA from digestion in the endosome. Salt which is originally present within the polymersome core is released upon its dissolution, creating an osmotic shock that ruptures the endosomal membrane, releasing the polyplex into the cytosol. Here, the pH is again at neutral and the polymer is deprotonated, releasing the DNA for transfer into the nucleus and subsequent gene expression [Lomas et al. 2007].

4.6.7.4   Drug Conjugates. Polymers can be conjugated to therapeutic agents in order to influence properties key to the drug's performance, such as plasma half-life, toxicity, and immunogenicity. For small drug molecules with appropriate functionality, it is possible to convert them into an ATRP polymerization initiator and then use them to grow a drug-functional polymer chain of defined length and polydispersity. This has been shown possible with aspirin, paracetamol, and dexamethasone. Alternatively, an ATRP polymer can be made with a suitable reactive group on the end of the chain for subsequent reaction with a therapeutic agent. Polymer modification of proteins has had a great impact, with the advent of PEGylated drugs such as polyethylene glycol-interferon (PEG-IFN). MPC, made by ATRP with a conjugating linker that inserts itself into the disulfide bridge of proteins, has been used to modify IFN-α and results in a conjugate that is stable against aggregation, resistant to IFN-antibody binding, extends plasma half-life in vivo, and yet maintains activity. Miyamoto et al. have synthesized PMPC by a living radical technique, then conjugated this to the enzyme papain, and demonstrated that the conjugated entity is more stable than the native enzyme [Miyamoto et al. 2004]. MPC-*co*-BMA-*co*-*p*-nitrophenylcarbonyl-oxyethyl methacrylate has been used as a backbone for the conjugation of IL-2 [Chiba et al. 2007b]. The BMA acts as a hydrophobic domain in water that can carry PTX or cycolsporin A; the drug loaded conjugates were able to suppress the proliferation of cell lines such as activated lymphocytes that overexpress high affinity IL-2 receptors. A similar polymer in which the BMA was replaced with the cationic monomer DMA was conjugated to hepatitis B surface antigen and condensed with plasmid DNA as a system for specific transfer of genes to human hepatocytes [Chiba et al. 2007a]. Gene expression was demonstrated in vitro using HepG-2 cells, and nonadverse side effects were observed in vivo in mice.

4.6.8   Emerging Applications

There are a plethora of reports of the use of PC polymers in a wide range of additional applications, many focused on emerging area of cell scaffolding and tissue engineering. There are, however, a number of applications that take advantage of other properties inherent in PC polymer. Given that desirable properties for a wound dressing include moisture retention at the wound site, nonadhesive properties to the healing tissue, and a general biological inertness in order not to aggravate the inflamed region, it is surprising there have been so few instances of the use of PC polymers in this application. MPC-*co*-LMA has been applied as a coating to a conventional PU dressing and has shown potential to provide an inert wound-healing environment in a rat skin wound model [Katakura et al. 2005]. Indeed, the moisturizing capability of MPC polymers is the basis of another of its commercial uses: as a cosmetic preparation. PMPC is the basis of a skin cream formulation available in Japan; it has been shown to protect the barrier properties of the stratum corneum by preventing disruption of the interlamellar bilayer that can result from excessive skin hydration [Lee 2004]. PMPC has also been combined with hyaluronic acid and studied in vitro and in vivo using a novel Raman microspectroscopic method and has shown to be effective as a moisturizer for cosmetic and dermatological uses [Chrit et al. 2007]. Moreover, MPC-*co*-BMA has been shown to be useful for inhibiting the skin permeation and antibacterial effect of parabens, probably as a result of the higher solubility in water and a lower partition to the skin and bacterial membranes of parabens by addition of the polymer [Hasegawa et al. 2005].

Cellular interactions have been shown to be reduced by the presence of a PC-based coating. This, however, could be advantageous in certain cases. For instance, the growth of antibody-secreting hybridomas requires special conditions such as serum-free defined media containing growth factors and vitamins. The surface on which these cells can proliferate has equally been shown to play an important role. This prompted a study to determine whether well-established hybridoma cell lines were able to proliferate and produce measurable amounts of monoclonal antibodies when grown on PC polymer-coated surfaces [Montano et al. 2005]. The results demonstrated that cell culture plates coated with an MPC copolymer were able to perform better than commercially available plates. These observations suggest that PC polymers could be used as an alternative, efficient surface coating to grow hybridoma cell lines and allow detectable antibody secretion. Other studies have aimed to modify the PC polymer in order to affect a specific cellular interaction. The MPC-modified hollow fiber membranes described in Section 4.6.6 have been taken a stage further and treated in such a way as to modify the surfaces asymmetrically, producing an MPC-rich nonfouling surface on the outside, and a charged, cell-friendly MPC surface on the inside [Ho Ye et al. 2006]. This has allowed hepatocytes to be cultured on the inner surface that show rounded morphology with higher functional expression in terms of urea and albumin synthesis; this is proposed as the basis of a liver assist bioreactor. Others have taken cellulose acetate membranes coated with MPC polymers as a system to encapsulate pancreatic islets [Yang et al. 2004]. The membranes provide a biocompatible immunobarrier while allowing passage of nutrients and small molecules such as glucose and insulin.

More complex biocompatible film formats that utilize PC polymers in conjunction with carbohydrate polymers are appearing in the literature. Meng et al. have combined PC and chitosan to enhance the biological properties of the material, and showed improved cell-material interactions with human umbilical vein endothelial cells with the hybrid polymer [Meng et al. 2007]. Others have described a layer-by-layer assembly of PC-modified chitosan and hyaluronan (HA), which are stable over a range of pH and possess high water content [Kujawa et al. 2007]. Iwasaki et al. have described MPC-BMA copolymer containing carbohydrate side chains that form surfaces suitable for biorecognition of HepG2 liver cancer cells and preservation of cell function [Iwasaki et al. 2007b].

4.7 SUMMARY

In recent years, perhaps the most commonly used polymer system investigated for the improvement of biological interactions is PEG. There is, however, a significant body of work, much of which has been reviewed in this chapter, that demonstrate the polymer systems based upon phospholipid biomimicry of the cell membrane are exceptional biomaterials. The ability of a PC polymer to suppress an adverse biological reaction stems from the fundamental ability to interact reversibly with proteins without inducing change in their conformation. This underlying property will provide the platform for future development of technologies that can be tuned to elicit a specific response by inclusion of a defined biological moiety, such as a ligand to a particular receptor, in order to target the system to a defined cell population. Such tunable biointerfaces comprising MPC nanoparticulates and polymer brushes form the subject of a recent review [Watanabe & Ishihara 2008].

The potential of these materials is already somewhat reflected in the commercialization of these systems, with significant in-market sales based upon stent and extracorporeal circuit coatings, contact lenses, surface treatments for various urological and other implants, and cosmetic cream formulations. The focus is, however, shifting with the growing interest in converging technologies. Drug-eluting stents are already enjoying the benefits that PC systems can offer in terms of biological properties and control of drug elution, and there are myriad of other combination devices to which this success could apply. Moreover, the number of reports on these polymers as drug delivery systems is increasing, and perhaps their more widespread application will follow as more information on the performance and fate of the polymer in vivo is generated.

EXERCISES/QUESTIONS FOR CHAPTER 4

1. Describe the rationale in the use of phosphorylcholine-mimicking biomaterials.
2. Provide the scheme of polymerization reaction for 2-methacryloyloxyethyl phosphorylcholine monomers and identify at least one typical mechanism of synthesis.
3. What is the main advantage in the use of phosphorylcholine-mimicking hydrogels?
4. Critically discuss the performance of phosphorylcholine-mimicking hydrogels in a typical clinical application.
5. Present the most relevant findings explaining the biocompatible character of medical devices based on phosphorylcholine-mimicking hydrogels or coatings.
6. List the main clinical applications where phosphorylcholine-mimicking polymers have been adopted as either bulk or coating biomaterials.
7. Provide a critical assessment of the ability of phosphorylcholine-mimicking polymers in enhancing tissue integration.
8. Discuss the physicochemical properties of the phosphorylcholine-mimicking polymers that in your view favor drug delivery.
9. Provide at least two examples of engineering of phosphorylcholine-mimicking polymers and their advantage in the relative clinical applications.
10. Critically discuss the potential use and limitations of phosphorylcholine-mimicking polymers in regenerative medicine.

REFERENCES

Abizaid A, Lansky AJ, Fitzgerald PJ, Tanajura LF, Feres F, Staico R, Mattos L, Abizaid A, Chaves A, Centemero M, Sousa AG, Sousa JE, Zaugg MJ, Schwartz LB (2007) Percutaneous coronary revascularization using a trilayer metal phosphorylcholine-coated zotarolimus-eluting stent. Am J Cardiol 99(10): 1403–1408.

Abraham S, Brahim S, Guiseppi-Elie A (2004) Molecularly engineered hydrogels for implant biocompatibility. Conf Proc IEEE Eng Med Biol Soc 7: 5036–5039.

Abraham S, Brahim S, Ishihara K, Guiseppi-Elie A (2005) Molecularly engineered p(HEMA)-based hydrogels for implant biochip biocompatibility. Biomaterials 26(23): 4767–4778.

Airoldi F, Di Mario C, Ribichini F, Presbitero P, Sganzerla P, Ferrero V, Vassanelli C, Briguori C, Carlino M, Montorfano M, Biondi-Zoccai GG, Chieffo A, Ferrari A, Colombo A (2005) 17-beta-estradiol eluting stent versus phosphorylcholine-coated stent for the treatment of native coronary artery disease. Am J Cardiol 96(5): 664–667.

Andrews CS, Denyer SP, Hall B, Hanlon GW, Lloyd AW (2001) A comparison of bacterial adhesion to standard HEMA and novel biomimetic soft contact lenses. Biomaterials 22: 3225–3233.

Armstrong J, Gunn J, Arnold N, Malik N, Chan KH, Vick T, Stratford P, Cumberland DC, Holt CM (2002) Angiopeptin-eluting stents: observations in human vessels and pig coronary arteries. J Invasive Cardiol 14(5): 230–238.

Bakhai A, Booth J, Delahunty N, Nugara F, Clayton T, McNeill J, Davies SW, Cumberland DC, Stables RH, SV Stent Investigators. (2005). The SV stent study: a prospective, multicentre, angiographic evaluation of the BiodivYsio phosphorylcholine coated small vessel stent in small coronary vessels. Int J Cardiol 102(1): 95–102.

Berrocal DH, Gonzalez GE, Morales C, Gelpi RJ, Grinfeld LR (2006) Irinotecan-eluting stents inhibited neointimal proliferation in hypercholesterolemic rabbit aortas. Catheter Cardiovasc Interv 68(1): 89–96.

Berry JA, Biedlingmaier JF, Whelan PJ (2000) In vitro resistance to bacterial biofilm formation on coated fluoroplastic tympanostomy tubes. Otolaryngol Head Neck Surg 123(3): 246–251.

Bi H, Zhong W, Meng S, Kong J, Yang P, Liu B (2006) Construction of a biomimetic surface on microfluidic chips for biofouling resistance. Anal Chem 78(10): 3399–3405.

Bird RL, Hall B, Hobbs KE, Chapman D (1989) New haemocompatible polymers assessed by thrombelastography. J Biomed Eng 11(3): 231–234.

Boland JL, Corbeij HA, Van Der Giessen W, Seabra-Gomes R, Suryapranata H, Wijns W, Hanet C, Suttorp MJ, Buller C, Bonnier JJ, Colombo A, Van Birgelen C, Pieper M, Mangioni JA, Londero H, Carere RG, Hamm CW, Bonan R, Bartorelli A, Kyriakides ZS, Chauhan A, Rothman M, Grinfeld L, Oosterwijk C, Serruys PW, Cumberland DC (2000) Multicenter evaluation of the phosphorylcholine-coated biodivYsio stent in short de novo coronary lesions: the SOPHOS study. Int J Cardiovasc Intervent 3(4): 215–225.

Burke SE, Kuntz RE, Schwartz LB (2006) Zotarolimus (ABT-578) eluting stents. Adv Drug Deliv Rev 58(3): 437–446.

Campbell EJ, O'Byrne V, Stratford PW, Quirk I, Vick TA, Wiles MC, Yianni YP (1994) Biocompatible surfaces using methacryloylphosphorylcholine laurylmethacrylate copolymer. ASAIO J 40(3): M853–M857.

Castelletto V, Hamley IW, Ma Y, Bories-Azeau X, Armes SP, Lewis AL (2004) Microstructure and physical properties of a pH-responsive gel based on a novel biocompatible ABA-type triblock copolymer. Langmuir 20(10): 4306–4309.

Chae KH, Jang YM, Kim YH, Sohn OJ, Rhee JI (2007) Anti-fouling epoxy coatings for optical biosensor application based on phosphorylcholine. Sens Actuators B Chem 124(1): 153–160.

Chan KH, Armstrong J, Withers S, Malik N, Cumberland DC, Gunn J, Holt CM (2007) Vascular delivery of c-myc antisense from cationically modified phosphorylcholine coated stents. Biomaterials 28(6): 1218–1224.

Chen C, Lumsden AB, Ofenloch JC, Noe B, Campbell EJ, Stratford PW, Yianni YP, Taylor AS, Hanson SR (1997) Phosphorylcholine coating of ePTFE grafts reduces neointimal hyperplasia in canine model. Ann Vasc Surg 11(1): 74–79.

Chen C, Ofenloch JC, Yianni YP, Hanson SR, Lumsden AB (1998) Phosphorylcholine coating of ePTFE reduces platelet deposition and neointimal hyperplasia in arteriovenous grafts. J Surg Res 77(2): 119–125.

Chen CY, Tamiya E, Ishihara K, Kosugi Y, Su YC, Nakabayashi N, Karube I (1992) A biocompatible needle-type glucose sensor based on platinum-electroplated carbon electrode. Appl Biochem Biotechnol 36(3): 211–226.

Chen CY, Ishihara K, Nakabayashi N, Tamiya E, Karube I (1999) Multifunctional biocompatible membrane and its application to fabricate a miniaturized glucose sensor with potential for use in vivo. Biomed Microdevices 1(2): 155–166.

Chevallier P, Janvier R, Mantovani D, Laroche G (2005) In vitro biological performances of phosphorylcholine-grafted ePTFE prostheses through RFGD plasma techniques. Macromol Biosci 5(9): 829–839.

Chiba N, Ueda M, Shimada T, Jinno H, Watanabe J, Ishihara K, Kitajima M (2007a) Development of gene vectors for pinpoint targeting to human hepatocytes by cationically modified polymer complexes. Eur Surg Res 39(1): 23–34.

Chiba N, Ueda M, Shimada T, Jinno H, Watanabe J, Ishihara K, Kitajima M (2007b) Novel immunosuppressant agents targeting activated lymphocytes by biocompatible MPC polymer conjugated with interleukin-2. Eur Surg Res 39(2): 103–110.

Chim YTA, Lam JKW, Ma Y, Armes SP, Lewis AL, Roberts CJ, Stolnik S, Tendler SJB, Davies MC (2005) Structural study of DNA condensation induced by novel phosphorylcholine-based copolymers for gene delivery and relevance to DNA protection. Langmuir 21(8): 3591–3598.

Chrit L, Bastien P, Biatry B, Simonnet JT, Potter A, Minondo AM, Flament F, Bazin R, Sockalingum GD, Leroy F, Manfait M, Hadjur C (2007) In vitro and in vivo confocal Raman study of human skin hydration: assessment of a new moisturizing agent, pMPC. Biopolymers 85(4): 359–369.

Chronos NAF, Campbell EJ, et al. (1994) Improved haemocompatibility of artificial surfaces can be achieved by phosphporylcholine coating: a human ex vivo flowing blood model. Eur Heart J 15: 312 (P1645).

Clarke S, Davies MC, Roberts CJ, Tendler SJB, Williams PM, O'Byrne V, Lewis AL, Russell J (2000) Surface mobility of 2-methacryloyloxyethyl phosphorylcholine-co-lauryl methacrylate polymers. Langmuir 16: 5116–5122.

Clarke S, Davies MC, Roberts CJ, Tendler SJB, Williams PM, Lewis AL, O'Byrne V (2001) Atomic force microscope and surface plasmon resonance investigation of polymer blends of poly([2-(methacryloyloxy)ethyl]phosphorylcholine-co-lauryl methacrylate) and poly(lauryl methacrylate). Macromolecules 34: 4166–4172.

Collingwood R, Gibson L, Sedlik S, Virmani R, Carter AJ (2005) Stent-based delivery of ABT-578 via a phosphorylcholine surface coating reduces neointimal formation in the porcine coronary model. Catheter Cardiovasc Interv 65(2): 227–232.

Court JL, Redman RP, Wang JH, Leppard SW, O'Byrne VJ, Small SA, Lewis AL, Jones SA, Stratford PW (2001) A novel phosphorylcholine-coated contact lens for extended wear use. Biomaterials 22(24): 3261–3272.

DeFife KM, Yun JK, Azeez A, Stack S, Ishihara K, Nakabayashi N, Colton E, Anderson JM (1995) Adhesion and cytokine production by monocytes on poly(2-methacryloyloxyethyl phosphorylcholine-co-alkyl methacrylate)-coated polymers. J Biomed Mater Res 29(4): 431–439.

De Somer F, François K, van Oeveren W, Poelaert J, De Wolf D, Ebels T, Van Nooten G (2000) Phosphorylcholine coating of extracorporeal circuits provides natural protection against blood activation by the material surface. Eur J Cardiothorac Surg 18(5): 602–606.

Du J, Tang Y, Lewis AL, Armes SP (2005) pH-Sensitive vesicles based on a biocompatible zwitterionic diblock copolymer. J Am Chem Soc 127(51): 17982–17983.

Dudley LY, Stratford PW, Aktar S, Hawes C, Reuben B, Perl O, Reed IM (1993) Coatings for the prevention of fouling of microfiltration membranes. Trans IChemE 71(A): 327–328.

Durrani AA, Hayward JA, Chapman D (1986) Biomembranes as models for polymer surfaces: II. The syntheses of reactive species for covalent coupling of phosphorylcholine to polymer surfaces. Biomaterials 7(2): 121–125.

Fajadet J, Wijns W, Laarman GJ, Kuck KH, Ormiston J, Münzel T, Popma JJ, Fitzgerald PJ, Bonan R, Kuntz RE, ENDEAVOR II Investigators. (2006) Randomized, double-blind, multicenter study of the Endeavor zotarolimus-eluting phosphorylcholine-encapsulated stent for treatment of native coronary artery lesions: clinical and angiographic results of the ENDEAVOR II trial. Circulation 114(8): 798–806.

Fajadet J, Wijns W, Laarman GJ, Kuck KH, Ormiston J, Münzel T, Popma JJ, Fitzgerald PJ, Bonan R, Kuntz RE, ENDEAVOR II Investigators. (2007) Randomized, double-blind, multicenter study of the Endeavor zotarolimus-eluting phosphorylcholine-encapsulated stent for treatment of native coronary artery lesions. Clinical and angiographic results of the ENDEAVOR II Trial. Minerva Cardioangiol 55(1): 1–18.

Fan D, Jia Z, Yan X, Liu X, Dong W, Sun F, Ji J, Xu J, Ren K, Chen W, Shen J, Qiu H, Gao R (2007) Pilot study of a cell membrane like biomimetic drug-eluting coronary stent. Sheng Wu Yi Xue Gong Cheng Xue Za Zhi 24(3): 599–602.

Feng W, Brash JL, Zhu S (2006) Non-biofouling materials prepared by atom transfer radical polymerization grafting of 2-methacryloxyethyl phosphorylcholine: separate effects of graft density and chain length on protein repulsion. Biomaterials 27(6): 847–855.

Fujii K, Matsumoto HN, Koyama Y, Iwasaki Y, Ishihara K, Takakuda K (2008) Prevention of biofilm formation with a coating of 2-methacryloyloxyethyl phosphorylcholine polymer. J Vet Med Sci 70(2): 167–173.

Gershlick A, Kandzari DE, Leon MB, Wijns W, Meredith IT, Fajadet J, Popma JJ, Fitzgerald PJ, Kuntz RE, ENDEAVOR Investigators. (2007) Zotarolimus-eluting stents in patients with native coronary artery disease: clinical and angiographic outcomes in 1317 patients. Am J Cardiol 100(8B): 45M–55M.

Giacomelli C, Le Men L, Borsali R, Lai-Kee-Him J, Brisson A, Armes SP, Lewis AL (2006) Phosphorylcholine-based pH-responsive diblock copolymer micelles as drug delivery vehicles: light scattering, electron microscopy, and fluorescence experiments. Biomacromolecules 7(3): 817–828.

Gobeil F, Juneau C, Plante S (2002) Thrombus formation on guide wires during routine PTCA procedures: a scanning electron microscopic evaluation. Can J Cardiol 18(3): 263–269.

Goda T, Ishihara K (2006) Soft contact lens biomaterials from bioinspired phospholipid polymers. Expert Rev Med Devices 3(2): 167–174.

Goda T, Watanabe J, Takai M, Ishihara K (2006) Water structure and improved mechanical properties of phospholipid polymer hydrogel with phosphorylcholine centered intermolecular cross-linker. Polymer 47(4): 1390–1396.

Goda T, Konno T, Takai M, Ishihara K (2007) Photoinduced phospholipid polymer grafting on Parylene film: advanced lubrication and antibiofouling properties. Colloids Surf B Biointerfaces 54(1): 67–73.

Goda T, Matsuno R, Konno T, Takai M, Ishihara K (2008) Photografting of 2-methacryloyloxyethyl phosphorylcholine from polydimethylsiloxane: tunable protein repellency and lubrication property. Colloids Surf B Biointerfaces 63(1): 64–72.

Goreish HH, Lewis AL, Rose S, Lloyd AW (2004) The effect of phosphorylcholine-coated materials on the inflammatory response and fibrous capsule formation: in vitro and in vivo observations. J Biomed Mater Res A 68(1): 1–9.

Goto Y, Matsuno R, Konno T, Takai M, Ishihara K (2008) Polymer nanoparticles covered with phosphorylcholine groups and immobilized with antibody for high-affinity separation of proteins. Biomacromolecules 9(3): 828–833.

Habib M, Brooks RA, Ireland DC, Lewis AL, Bonfield W (2003) Interaction of human osteoblasts with phosphorylcholine polymers. Eur Cell Mater 6(1): 35.

Hall B, Bird RR, Kojima M, Chapman D (1989) Biomembranes as models for polymer surfaces: V. Thrombelastographic studies of polymeric lipids and polyesters. Biomaterials 10(4): 219–224.

Han SH, Ahn TH, Kang WC, Oh KJ, Chung WJ, Shin MS, Koh KK, Choi IS, Shin EK (2006) The favorable clinical and angiographic outcomes of a high-dose dexamethasone-eluting stent: randomized controlled prospective study. Am Heart J 152(5): 887 e1–887 e7.

Han Y, Liang M, Kang J, Qi Y, Deng J, Xu K, Yan C (2007) Estrogen-eluting stent implantation inhibits neointimal formation and extracellular signal-regulated kinase activation. Catheter Cardiovasc Interv 70(5): 647–653.

Hasegawa T, Iwasaki Y, Ishihara K (2001) Preparation and performance of protein-adsorption-resistant asymmetric porous membrane composed of polysulfone/phospholipid polymer blend. Biomaterials 22(3): 243–251.

Hasegawa T, Iwasaki Y, Ishihara K (2002) Preparation of blood-compatible hollow fibers from a polymer alloy composed of polysulfone and 2-methacryloyloxyethyl phosphorylcholine polymer. J Biomed Mater Res 63(3): 333–341.

Hasegawa T, Kim S, Tsuchida M, Issiki Y, Kondo S, Sugibayashi K (2005) Decrease in skin permeation and antibacterial effect of parabens by a polymeric additive, poly(2-methacryloyloxyethyl phosphorylcholine-co-butylmetacrylate). Chem Pharm Bull (Tokyo) 53(3): 271–276.

Hayward JA, Chapman D (1984) Biomembrane surfaces as models for polymer design: the potential for haemocompatibility. Biomaterials 5(3): 135–142.

Hayward JA, Durrani AA, Lu Y, Clayton CR, Chapman D (1986a) Biomembranes as models for polymer surfaces: IV. ESCA analyses of a phosphorylcholine surface covalently bound to hydroxylated substrates. Biomaterials 7(4): 252–258.

Hayward JA, Durrani AA, Shelton CJ, Lee DC, Chapman D (1986b) Biomembranes as models for polymer surfaces. III. Characterization of a phosphorylcholine surface covalently bound to glass. Biomaterials 7(2): 126–131.

Hirota K, Murakami K, Nemoto K, Miyake Y (2005) Coating of a surface with 2-methacryloyloxyethyl phosphorylcholine (MPC) co-polymer significantly reduces retention of human pathogenic microorganisms. FEMS Microbiol Lett 248(1): 37–45.

Hoven VP, Srinanthakul M, Iwasaki Y, Iwata R, Kiatkamjornwong S (2007) Polymer brushes in nanopores surrounded by silicon-supported tris(trimethylsiloxy)silyl monolayers. J Colloid Interface Sci 314(2): 446–459.

Ho Ye S, Watanabe J, Takai M, Iwasaki Y, Ishihara K (2006) High functional hollow fiber membrane modified with phospholipid polymers for a liver assist bioreactor. Biomaterials 27(9): 1955–1962.

Hsiue GH, Lee SD, Chang PC, Kao CY (1998) Surface characterization and biological properties study of silicone rubber membrane grafted with phospholipid as biomaterial via plasma induced graft copolymerization. J Biomed Mater Res 42(1): 134–147.

Hsiue GH, Lo CL, Cheng CH, Lin CP, Huang CK, Chen HH (2007) Preparation and characterisation of poly(2-methacryloyloxyethyl phosphorylcholine)-block-poly(D,L-lactide) polymer nanoparticles. J Polym Sci A 45(4): 688–698.

Huang Y, Liu X, Wang L, Verbeken E, Li S, De Scheerder I (2003) Local methylprednisolone delivery using a BiodivYsio phosphorylcholine-coated drug-delivery stent reduces inflammation and neointimal hyperplasia in a porcine coronary stent model. Int J Cardiovasc Intervent 5(3): 166–171.

Ishihara K, Lifeng Y (2008) Graft copolymerization of 2-methacryloyloxyethyl phosphorylcholine to cellulose in homogeneous media using atom transfer radical polymerization for providing new hemocompatible coating materials. J Polym Sci A 46(10): 3306–3313.

Ishihara K, Ziats NP, Tierney BP, Nakabayashi N, Anderson JM (1991) Protein adsorption from human plasma is reduced on phospholipid polymers. J Biomed Mater Res 25(11): 1397–1407.

Ishihara K, Oshida H, Endo Y, Ueda T, Watanabe A, Nakabayashi N (1992) Hemocompatibility of human whole blood on polymers with a phospholipid polar group and its mechanism. J Biomed Mater Res 26(12): 1543–1552.

Ishihara K, Inoue H, Kurita K, Nakabayashi N (1994a) Selective adhesion of platelets on a polyion complex composed of phospholipid polymers containing sulfonate groups and quarternary ammonium groups. J Biomed Mater Res 28(11): 1347–1355.

Ishihara K, Tsuji T, Kurosaki T, Nakabayashi N (1994b) Hemocompatibility on graft copolymers composed of poly(2-methacryloyloxyethyl phosphorylcholine) side chain and poly(n-butyl methacrylate) backbone. J Biomed Mater Res 28(2): 225–232.

Ishihara K, Shibata N, Tanaka S, Iwasaki Y, Kurosaki T, Nakabayashi N (1996a) Improved blood compatibility of segmented polyurethane by polymeric additives having phospholipid polar group: II. Dispersion state of the polymeric additive and protein adsorption on the surface. J Biomed Mater Res 32(3): 401–408.

Ishihara K, Tanaka S, Furukawa N, Kurita K, Nakabayashi N (1996b) Improved blood compatibility of segmented polyurethanes by polymeric additives having phospholipid polar groups: I. Molecular design of polymeric additives and their functions. J Biomed Mater Res 32(3): 391–399.

Ishihara K, Nomura H, Mihara T, Kurita K, Iwasaki Y, Nakabayashi N (1998) Why do phospholipid polymers reduce protein adsorption? J Biomed Mater Res 39(2): 323–330.

Ishihara K, Fukumoto K, Iwasaki Y, Nakabayashi N (1999a) Modification of polysulfone with phospholipid polymer for improvement of the blood compatibility: Part 1. Surface characterization. Biomaterials 20(17): 1545–1551.

Ishihara K, Fukumoto K, Iwasaki Y, Nakabayashi N (1999b) Modification of polysulfone with phospholipid polymer for improvement of the blood compatibility: Part 2. Protein adsorption and platelet adhesion. Biomaterials 20(17): 1553–1559.

Ishihara K, Ishikawa E, Iwasaki Y, Nakabayashi N (1999c) Inhibition of fibroblast cell adhesion on substrate by coating with 2-methacryloyloxyethyl phosphorylcholine polymers. J Biomater Sci Polym Ed 10(10): 1047–1061.

Ishihara K, Fujita H, Yoneyama T, Iwasaki Y (2000) Antithrombogenic polymer alloy composed of 2-methacryloyloxyethyl phosphorylcholine polymer and segmented polyurethane. J Biomater Sci Polym Ed 11(11): 1183–1195.

Ishihara K, Hasegawa T, Watanabe J, Iwasaki Y (2002) Protein adsorption-resistant hollow fibers for blood purification. Artif Organs 26(12): 1014–1019.

Ishihara K, Nishiuchi D, Watanabe J, Iwasaki Y (2004) Polyethylene/phospholipid polymer alloy as an alternative to poly(vinylchloride)-based materials. Biomaterials 25(6): 1115–1122.

Ito T, Iwasaki Y, Narita T, Akiyoshi K, Ishihara K (2005) Cell separation in microcanal coated with electrically charged phospholipid polymers. Colloids Surf B Biointerfaces 41(2-3): 175–180.

Ito T, Watanabe J, Takai M, Konno T, Iwasaki Y, Ishihara K (2006) Dual mode bioreactions on polymer nanoparticles covered with phosphorylcholine group. Colloids Surf B Biointerfaces 50(1): 55–60.

Iwasaki Y, Ishihara K (2005) Phosphorylcholine-containing polymers for biomedical applications. Anal Bioanal Chem 381(3): 534–546.

Iwasaki Y, Kurita K, Ishihara K, Nakabayashi N (1994) Effect of methylene chain length in phospholipid moiety on blood compatibility of phospholipid polymers. J Biomater Sci Polym Ed 6(5): 447–461.

Iwasaki Y, Fujiike A, Kurita K, Ishihara K, Nakabayashi N (1996a) Protein adsorption and platelet adhesion on polymer surfaces having phospholipid polar group connected with oxyethylene chain. J Biomater Sci Polym Ed 8(2): 91–102.

Iwasaki Y, Kurita K, Ishihara K, Nakabayashi N (1996b) Effect of reduced protein adsorption on platelet adhesion at the phospholipid polymer surfaces. J Biomater Sci Polym Ed 8(2): 151–163.

Iwasaki Y, Sawada S, Ishihara K, Khang G, Lee HB (2002) Reduction of surface-induced inflammatory reaction on PLGA/MPC polymer blend. Biomaterials 23(18): 3897–3903.

Iwasaki Y, Yamasaki A, Ishihara K (2003) Platelet compatible blood filtration fabrics using a phosphorylcholine polymer having high surface mobility. Biomaterials 24(20): 3599–3604.

Iwasaki Y, Maie H, Akiyoshi K (2007a) Cell-specific delivery of polymeric nanoparticles to carbohydrate-tagging cells. Biomacromolecules 8(10): 3162–3168.

Iwasaki Y, Takami U, Shinohara Y, Kurita K, Akiyoshi K (2007b) Selective biorecognition and preservation of cell function on carbohydrate-immobilized phosphorylcholine polymers. Biomacromolecules 8(9): 2788–2794.

Iwasaki Y, Takamiya M, Iwata R, Yusa S, Akiyoshi K (2007c) Surface modification with well-defined biocompatible triblock copolymers: improvement of biointerfacial phenomena on a poly(dimethylsiloxane) surface. Colloids Surf B Biointerfaces 57(2): 226–236.

Iwata R, Satoh R, Iwasaki Y, Akiyoshi K (2008) Covalent immobilization of antibody fragments on well-defined polymer brushes via site-directed method. Colloids Surf B Biointerfaces 62(2): 288–298.

Jiang W, Fischer G, Girmay Y, Irgum K (2006) Zwitterionic stationary phase with covalently bonded phosphorylcholine type polymer grafts and its applicability to separation of peptides in the hydrophilic interaction liquid chromatography mode. J Chromatogr A 1127(1-2): 82–91.

Jin Q, Xu JP, Ji J, Shen JC (2008) Zwitterionic phosphorylcholine as a better ligand for stabilizing large biocompatible gold nanoparticles. Chem Commun (Camb) 26: 3058–3060.

Kadoma Y, Nakabayashi N, Masuhara E, Yamauchi J (1978) Synthesis and hemolysis test of the polymer containing phosphorylcholine groups. Kobunshi Ronbunshu 35(7): 423–427.

Kandzari DE, Leon MB (2006) Overview of pharmacology and clinical trials program with the zotarolimus-eluting endeavor stent. J Interv Cardiol 19(5): 405–413.

Kandzari DE, Leon MB, Popma JJ, Fitzgerald PJ, O'Shaughnessy C, Ball MW, Turco M, Applegate RJ, Gurbel PA, Midei MG, Badre SS, Mauri L, Thompson KP, LeNarz LA, Kuntz RE, ENDEAVOR III Investigators. (2006) Comparison of zotarolimus-eluting and sirolimus-eluting stents in patients with native coronary artery disease: a randomized controlled trial. J Am Coll Cardiol 48(12): 2440–2447.

Katakura O, Morimoto N, Iwasaki Y, Akiyoshi K, Kasugai S (2005) Evaluation of 2-methacryloyloxyethyl phosphorylcholine (MPC) polymer-coated dressing on surgical wounds. J Med Dent Sci 52(2): 115–121.

Kimura M, Fukumoto K, Watanabe J, Ishihara K (2004) Hydrogen-bonding-driven spontaneous gelation of water-soluble phospholipid polymers in aqueous medium. J Biomater Sci Polym Ed 15(5): 631–644.

Kimura M, Fukumoto K, Watanabe J, Takai M, Ishihara K (2005) Spontaneously forming hydrogel from water-soluble random- and block-type phospholipid polymers. Biomaterials 26(34): 6853–6862.

Kimura M, Konno T, Takai M, Ishiyama N, Moro T, Ishihara K (2007a) Prevention of tissue adhesion by spontaneously formed phospholipid polymer hydrogel. Key Eng Mat 342-343: 777–780.

Kimura M, Takai M, Ishihara K (2007b) Biocompatibility and drug release behavior of spontaneously formed phospholipid polymer hydrogels. J Biomed Mater Res A 80(1): 45–54.

Kimura M, Takai M, Ishihara K (2007c) Tissue-compatible and adhesive polyion complex hydrogels composed of amphiphilic phospholipid polymers. J Biomater Sci Polym Ed 18(5): 623–640.

Kiritoshi Y, Ishihara K (2002) Preparation of cross-linked biocompatoble poly(2-methacryloyloxyethyl phosphorylcholine) gel and its strange swelling behaviour in water/ethanol mixture. J Biomater Sci Polym Ed 13(2): 213–224.

Kitano H, Sudo K, Ichikawa K, Ide M, Ishihara K (2000) Raman spectroscopic study on the structure of water in aqueous polyelectrolyte solutions. J Phys Chem B 104(47): 11425–11429.

Kitano H, Imai M, Mori T, Gemmei-Ide M, Yokoyama Y, Ishihara K (2003) Structure of water in the vicinity of phospholipid analogue copolymers as studied by vibrational spectroscopy. Langmuir 19(24): 10260–10266.

Kitano H, Mori T, Takeuchi Y, Tada S, Gemmei-Ide M, Yokoyama Y, Tanaka M (2005) Structure of water incorporated in sulfobetaine polymer films as studied by ATR-FTIR. Macromol Biosci 5(4): 314–321.

Kitano H, Takaha K, Gemmei-Ide M (2006) Raman spectroscopic study of the structure of water in aqueous solutions of amphoteric polymers. Phys Chem Chem Phys 8(10): 1178–1185.

Kobayashi K, Ohuchi K, Hoshi H, Morimoto N, Iwasaki Y, Takatani S (2005) Segmented poly-urethane modified by photopolymerization and cross-linking with 2-methacryloyloxyethyl phosphorylcholine polymer for blood-contacting surfaces of ventricular assist devices. J Artif Organs 8(4): 237–244.

Kobayashi M, Terayama Y, Hosaka N, Kaido M, Suzuki A, Yamada N, Torikai N, Ishihara K, Takahara A (2007) Friction behavior of high denisty poly(2-methacryloyloxyethyl phosphor-ylcholine) brush in aqueous media. Soft Matter 3(6): 740–746.

Konno T, Ishihara K (2007) Temporal and spatially controllable cell encapsulation using a water-soluble phospholipid polymer with phenylboronic acid moiety. Biomaterials 28(10): 1770–1777.

Konno T, Watanabe J, Ishihara K (2003) Enhanced solubility of paclitaxel using water-soluble and biocompatible 2-methacryloyloxyethyl phosphorylcholine polymers. J Biomed Mater Res A 65(2): 209–214.

Konno T, Hasuda H, Ishihara K, Ito Y (2005) Photo-immobilization of a phospholipid polymer for surface modification. Biomaterials 26(12): 1381–1388.

Konno T, Ito T, Takai M, Ishihara K (2006) Enzymatic photochemical sensing using luciferase-immobilized polymer nanoparticles covered with artificial cell membrane. J Biomater Sci Polym Ed 17(12): 1347–1357.

Korematsu A, Takemoto Y, Nakaya T, Inoue H (2002) Synthesis, characterization and platelet adhesion of segmented polyurethanes grafted phospholipid analogous vinyl monomer on surface. Biomaterials 23(1): 263–271.

Kudo H, Sawada T, Kazawa E, Yoshida H, Iwasaki Y, Mitsubayashi K (2006) A flexible and wearable glucose sensor based on functional polymers with soft-MEMS techniques. Biosens Bioelectron 22(4): 558–562.

Kudo H, Yagi T, et al. (2008) Glucose sensor using a phospholipid polymer-based enzyme immobilization method. Anal Bioanal Chem 391(4): 1269–1274.

Kuiper KK, Robinson KA, Chronos NA, Cui J, Palmer SJ, Nordrehaug JE (1998) Phosphorylcholine-coated metallic stents in rabbit iliac and porcine coronary arteries. Scand Cardiovasc J 32(5): 261–268.

Kujawa P, Schmauch G, Viitala T, Badia A, Winnik FM (2007) Construction of viscoelastic biocompatible films via the layer-by-layer assembly of hyaluronan and phosphorylcholine-modified chitosan. Biomacromolecules 8(10): 3169–3176.

Kwok OH, Chow WH, Law TC, Chiu A, Ng W, Lam WF, Hong MK, Popma JJ (2005) First human experience with angiopeptin-eluting stent: a quantitative coronary angiography and three-dimensional intravascular ultrasound study. Catheter Cardiovasc Interv 66(4): 541–546.

Kyomoto M, Iwasaki Y, Moro T, Konno T, Miyaji F, Kawaguchi H, Takatori Y, Nakamura K, Ishihara K (2007a) High lubricious surface of cobalt-chromium-molybdenum alloy prepared by grafting poly(2-methacryloyloxyethyl phosphorylcholine). Biomaterials 28(20): 3121–3130.

Kyomoto M, Moro T, Konno T, Takadama H, Yamawaki N, Kawaguchi H, Takatori Y, Nakamura K, Ishihara K (2007b) Enhanced wear resistance of modified cross-linked polyethylene by grafting with poly(2-methacryloyloxyethyl phosphorylcholine). J Biomed Mater Res A 82(1): 10–17.

Kyomoto M, Moro T, Miyaji F, Konno T, Hashimoto M, Kawaguchi H, Takatori Y, Nakamura K, Ishihara K (2008) Enhanced wear resistance of orthopaedic bearing due to the cross-linking of poly(MPC) graft chains induced by gamma-ray irradiation. J Biomed Mater Res B Appl Biomater 84(2): 320–327.

Lam JK, Ma Y, Armes SP, Lewis AL, Baldwin T, Stolnik S (2004) Phosphorylcholine-polycation diblock copolymers as synthetic vectors for gene delivery. J Control Release 100(2): 293–312.

Lee AR (2004) Phospholipid polymer, 2-methacryloyloxyethyl phosphorylcholine and its skin barrier function. Arch Pharm Res 27(11): 1177–1182.

Lewis AL (2000) Phosphorylcholine-based polymers and their use in the prevention of biofouling. Colloids Surf B Biointerfaces 18(3-4): 261–275.

Lewis AL (2004) Phosphorylcholine (PC) technology. In: Encyclopedia of Biomaterials and Biomedical Engineering, ed. Oona Schmid, 1198–1211. New York: Marcel Dekker.

Lewis AL (2006) PC technology as a platform for drug delivery: from combination to conjugation. Expert Opin Drug Deliv 3(2): 289–298.

Lewis AL, Stratford PW (2002) Phosphorylcholine-coated stents. J Long Term Eff Med Implants 12(4): 231–250.

Lewis AL, Vick TA (2001) Site-specific drug delivery from coronary stents. Drug Deliv Syst Sci 1(3): 65–71.

Lewis AL, Hughes PD, Kirkwood LC, Leppard SW, Redman RP, Tolhurst LA, Stratford PW (2000) Synthesis and characterisation of phosphorylcholine-based polymers useful for coating blood filtration devices. Biomaterials 21(18): 1847–1859.

Lewis AL, Cumming ZL, Goreish HH, Kirkwood LC, Tolhurst LA, Stratford PW (2001a) Cross-linkable coatings from phosphorylcholine-based polymers. Biomaterials 22(2): 99–111.

Lewis AL, Vick TA, Collias AC, Hughes LG, Palmer RR, Leppard SW, Furze JD, Taylor AS, Stratford PW (2001b) Phosphorylcholine-based polymer coatings for stent drug delivery. J Mater Sci Mater Med 12(10–12): 865–870.

Lewis AL, Furze JD, Small S, Robertson JD, Higgins BJ, Taylor S, Ricci DR (2002a) Long-term stability of a coronary stent coating post-implantation. J Biomed Mater Res 63(6): 699–705.

Lewis AL, Tolhurst LA, Stratford PW (2002b) Analysis of a phosphorylcholine-based polymer coating on a coronary stent pre- and post-implantation. Biomaterials 23(7): 1697–1706.

Lewis AL, Freeman RN, Redman RP, Tolhurst LA, Kirkwood LC, Grey DM, Vick TA (2003) Wettable phosphorylcholine-containing polymers useful in blood filtration. J Mater Sci Mater Med 14(1): 39–45.

Lewis AL, Berwick J, Davies MC, Roberts CJ, Wang JH, Small S, Dunn A, O'Byrne V, Redman RP, Jones SA (2004a) Synthesis and characterisation of cationically modified phospholipid polymers. Biomaterials 25(15): 3099–3108.

Lewis AL, Willis SL, Small SA, Hunt SR, O'byrne V, Stratford PW (2004b) Drug loading and elution from a phosphorylcholine polymer-coated coronary stent does not affect long-term stability of the coating in vivo. Biomed Mater Eng 14(4): 355–370.

Li C, Buurma NJ, Haq I, Turner C, Armes SP, Castelletto V, Hamley IW, Lewis AL (2005a) Synthesis and characterization of biocompatible, thermoresponsive ABC and ABA triblock copolymer gelators. Langmuir 21(24): 11026–11033.

Li C, Tang Y, Armes SP, Morris CJ, Rose SF, Lloyd AW, Lewis AL (2005b) Synthesis and characterization of biocompatible thermo-responsive gelators based on ABA triblock copolymers. Biomacromolecules 6(2): 994–999.

Li C, Madsen J, Armes SP, Lewis AL (2006) A new class of biochemically degradable, stimulus-responsive triblock copolymer gelators. Angew Chem Int Ed Engl 45(21): 3510–3513.

Li Y, Tang Y, Narain R, Lewis AL, Armes SP (2005) Biomimetic stimulus-responsive star diblock gelators. Langmuir 21(22): 9946–9954.

Licciardi M, Tang Y, Billingham NC, Armes SP, Lewis AL (2005) Synthesis of novel folic acid-functionalized biocompatible block copolymers by atom transfer radical polymerization for gene delivery and encapsulation of hydrophobic drugs. Biomacromolecules 6(2): 1085–1096.

Licciardi M, Giammona G, Du J, Armes SP, Tang Y, Lewis AL (2006) New folate-functionalised biocompatible block copolymer micelles as potential anti-cancer drug delivery systems. Polymer 47: 2946–2955.

Licciardi M, Craparo EF, Giammona G, Armes SP, Tang Y, Lewis AL (2008) in vitro biological evaluation of folate-functionalized block copolymer micelles for selective anti-cancer drug delivery. Macromol Biosci 8(7): 615–626.

Lim KS (1999) Cell and protein adhesion studies in glaucoma drainage device development. The AGFID project team. Br J Ophthalmol 83(10): 1168–1171.

Lim KS (2003) Corneal endothelial cell damage from glaucoma drainage device materials. Cornea 22(4): 352–354.

Lloyd AW, Dropcova S, Faragher RG, Gard PR, Hanlon GW, Mikhalovsky SV, Olliff CJ, Denyer SP, Letko E, Filipec M (1999) The development of in vitro biocompatibility tests for the evaluation of intraocular biomaterials. J Mater Sci Mater Med 10(10/11): 621–627.

Lloyd AW, Faragher RGA, Wassall M, Rhys-Williams W, Wong L, Hughes JE, Hanlon GW (2000) Assessing the in vitro cell-based ocular compatibility of contact lens materials. Cont Lens Anterior Eye 23: 119–123.

Lobb EJ, Ma I, Billingham NC, Armes SP, Lewis AL (2001) Facile synthesis of well-defined, biocompatible phosphorylcholine-based methacrylate copolymers via atom transfer radical polymerization at 20 degrees C. J Am Chem Soc 123(32): 7913–7914.

Lomas H, Canton I, MacNeil S, Du J, Armes SP, Ryan AJ, Lewis AL, Battaglia G (2007) Biomimetic pH sensitive polymersomes for efficient DNA encapsulation and delivery. Adv Mater 19: 4239–4243.

Long SF, Clarke S, Davies MC, Lewis AL, Hanlon GW, Lloyd AW (2003) Controlled biological response on blends of a phosphorylcholine-based copolymer with poly(butyl methacrylate). Biomaterials 24(23): 4115–4121.

Ma IY, Lobb EJ, Billingham NC, Armes SP, Lewis AL, Lloyd AW, Salvage JP (2002) Synthesis of biocompatible polymers 1. Homopolymerisation of 2-methacryloyloxyethyl phosphorylcholine via ATRP in protic solvents: an optimisation study. Macromolecules 35(25): 9306–9314.

Ma YH, Tang YQ, Billingham NC, Armes SP, Lewis AL, Lloyd AW, Salvage JP (2003) Well-defined biocompatibles block copolymers via atom transfer radical polymerization of 2-methacryloyloxyethyl phosphorylcholine in protic media. Macromolecules 36: 3475–3484.

Madsen J, Armes SP, Bertal K, Lomas H, Macneil S, Lewis AL (2008) Biocompatible wound dressings based on chemically degradable triblock copolymer hydrogels. Biomacromolecules 9(8): 2265–2275.

Malik N, Gunn J, Shepherd L, Crossman DC, Cumberland DC, Holt CM (2001) Phosphorylcholine-coated stents in porcine coronary arteries: in vivo assessment of biocompatibility. J Invasive Cardiol 13(3): 193–201.

Matsuda Y, Kobayashi M, Annaka M, Ishihara K, Takahara A. (2008) UCST-type cononsolvency behavior of poly(2-methacryloyloxyethyl phosphorylcholine) in the mixture of water and ethanol. Polym J 40(5): 479–483.

McNair AM (1996) Using hydrogel polymers for drug delivery. Med Device Technol 7(10): 16–22.

Meng S, Liu Z, Zhong W, Wang Q, Du Q (2007) Phosphorylcholine modified chitosan: appetent and safe material for cells. Carbohydr Poly 70(1): 82–88.

Miyamoto D, Watanabe J, Ishihara K (2004) Effect of water-soluble phospholipid polymers conjugated with papain on the enzymatic stability. Biomaterials 25(1): 71–76.

Montano X, Lewis AL, Leppard SW, Lichtenstein C (2005) Phosphorylcholine is favorable for antibody production from hybridoma cells. Biotechnol Bioeng 90(6): 770–774.

Morisaku T, Ikehara T, Watanabe J, Takai M, Ishihara K (2005) Design of biocompatible hydrogels with attention to structure of water surrounding polar groups in polymer chains. Trans Mater Res Soc Jpn 30(3): 835–838.

Moro T, Takatori Y, Ishihara K, Konno T, Takigawa Y, Matsushita T, Chung UI, Nakamura K, Kawaguchi H (2004) Surface grafting of artificial joints with a biocompatible polymer for preventing periprosthetic osteolysis. Nat Mater 3(11): 829–836.

Moro T, Takatori Y, Ishihara K, Nakamura K, Kawaguchi H (2006) 2006 Frank Stinchfield Award: grafting of biocompatible polymer for longevity of artificial hip joints. Clin Orthop Relat Res 453: 58–63.

Mu QS, Ma YH, Lewis AL, Armes SP, Lu JR (2004) Complexation of DNA with biocompatible diblock copolymers. Prog Colloid Polym Sci 128: 199–202.

Mu QS, Lu JR, Ma YH, Paz de Banez MV, Robinson KL, Armes SP, Lewis AL, Thomas RK (2006) Neutron reflection study of a water-soluble biocompatible diblock copolymer adsorbed at the air/water interface: the effects of pH and polymer concentration. Langmuir 22(14): 6153–6160.

Murphy EF, Keddie JL, Lu JR, Brewer J, Russell J (1999) The reduced adsorption of lysozyme at the phosphorylcholine incorporated polymer/aqueous solution interface studied by spectroscopic ellipsometry. Biomaterials 20(16): 1501–1511.

Murphy EF, Lu JR, Lewis AL, Brewer J, Russell J, Stratford P (2000) Characterisation of protein adsorption at the phosphorylcholine incorporated polymer-water interface. Macromolecules 33: 4545–4554.

Myers GJ, Johnstone DR, Swyer WJ, McTeer S, Maxwell SL, Squires C, Ditmore SN, Power CV, Mitchell LB, Ditmore JE, Aniuk LD, Hirsch GM, Buth KJ (2003) Evaluation of Mimesys phosphorylcholine (PC)-coated oxygenators during cardiopulmonary bypass in adults. J Extra Corpor Technol 35(1): 6–12.

Nakabayashi N, Iwasaki Y (2004) Copolymers of 2-methacryloyloxyethyl phosphorylcholine (MPC) as biomaterials. Biomed Mater Eng 14(4): 345–354.

Nakabayashi N, Williams DF (2003) Preparation of non-thrombogenic materials using 2-methacryloyloxyethyl phosphorylcholine. Biomaterials 24(13): 2431–2435.

Nakaya T, Li Y-J (1999) Phospholipid polymers. Prog Polym Sci 24(1): 143–181.

Nam K, Watanabe J, Ishihara K (2004a) The characteristics of spontaneously forming physically cross-linked hydrogels composed of two water-soluble phospholipid polymers for oral drug delivery carrier I: hydrogel dissolution and insulin release under neutral pH condition. Eur J Pharm Sci 23(3): 261–270.

Nam K, Watanabe J, Ishihara K (2004b) Modeling of swelling and drug release behavior of spontaneously forming hydrogels composed of phospholipid polymers. Int J Pharm 275(1-2): 259–269.

Nam KW, Watanabe J, Ishihara K (2002a) Characterization of the spontaneously forming hydrogels composed of water-soluble phospholipid polymers. Biomacromolecules 3(1): 100–105.

Nam KW, Watanabe J, Ishihara K (2002b) pH-Modulated release of insulin entrapped in a spontaneously formed hydrogel system composed of two water-soluble phospholipid polymers. J Biomater Sci Polym Ed 13(11): 1259–1269.

Narain R, Housni A, Lane L (2006) Modification of carboxyl-functionalized single-walled carbon nanotubes with biocompatible, water-soluble phosphorylcholine and suger-based polymers: bioinspired nanorods. J Polym Sci A 44(22): 6558–6568.

Navarro-Villoslada F, Orellana G, Moreno-Bondi MC, Vick T, Driver M, Hildebrand G, Liefeith K (2001) Fiber-optic luminescent sensors with composite oxygen-sensitive layers and anti-biofouling coatings. Anal Chem 73(21): 5150–5156.

New G, Moses JW, Roubin GS, Leon MB, Colombo A, Iyer SS, Tio FO, Mehran R, Kipshidze N (2002) Estrogen-eluting, phosphorylcholine-coated stent implantation is associated with reduced neointimal formation but no delay in vascular repair in a porcine coronary model. Catheter Cardiovasc Interv 57(2): 266–271.

Nishizawa K, Konno T, Takai M, Ishihara K (2008) Bioconjugated phospholipid polymer biointerface for enzyme-linked immunosorbent assay. Biomacromolecules 9(1): 403–407.

Okajima Y, Saika S, Sawa M (2005) Effects of surface modification on foreign body reaction of intraocular lenses. Nippon Ganka Gakkai Zasshi 109(5): 267–273.

Okajima Y, Kobayakawa S, Tsuji A, Tochikubo T (2006a) Biofilm formation by Staphylococcus epidermidis on intraocular lens material. Invest Ophthalmol Vis Sci 47(7): 2971–2975.

Okajima Y, Saika S, Sawa M (2006b) Effect of surface coating an acrylic intraocular lens with poly(2-methacryloyloxyethyl phosphorylcholine) polymer on lens epithelial cell line behavior. J Cataract Refract Surg 32(4): 666–671.

Palmer RR, Lewis AL, Kirkwood LC, Rose SF, Lloyd AW, Vick TA, Stratford PW (2004) Biological evaluation and drug delivery application of cationically modified phospholipid polymers. Biomaterials 25(19): 4785–4796.

Pappalardo F, Della Valle P, Crescenzi G, Corno C, Franco A, Torracca L, Alfieri O, Galli L, Zangrillo A, D'Angelo A (2006) Phosphorylcholine coating may limit thrombin formation during high-risk cardiac surgery: a randomized controlled trial. Ann Thorac Surg 81(3): 886–891.

Park J, Kurosawa S, Takai M, Ishihara K (2007) Antibody immobilization to phospholipid polymer layer on gold substrate of quartz crystal microbalance immunosensor. Colloids Surf B Biointerfaces 55(2): 164–172.

Park JW, Kurosawa S, Aizawa H, Goda Y, Takai M, Ishihara K (2006) Piezoelectric immunosensor for bisphenol A based on signal enhancing step with 2-methacrolyloxyethyl phosphorylcholine polymeric nanoparticle. Analyst 131(1): 155–162.

Parker AP, Reynolds PA, Lewis AL, Kirkwood L, Hughes LG (2005) Investigation into potential mechanisms promoting biocompatibility of polymeric biomaterials containing the phosphorylcholine moiety: a physicochemical and biological study. Colloids Surf B Biointerfaces 46(4): 204–217.

Patel JD, Iwasaki Y, Ishihara K, Anderson JM (2005) Phospholipid polymer surfaces reduce bacteria and leukocyte adhesion under dynamic flow conditions. J Biomed Mater Res A 73(3): 359–366.

Poilane N, Takai M, Ishihara K (2007) Well defined surface preparation with phospholipid polymers for highly sensitive immunoassays. Key Eng Mat 342-343: 889–892.

Radke PW, Griesenbach U, Kivela A, Vick T, Judd D, Munkonge F, Willis S, Geddes DM, Yla-Herttuala S, Alton EW (2005) Vascular oligonucleotide transfer facilitated by a polymer-coated stent. Hum Gene Ther 16(6): 734–740.

Ranucci M, Isgrò G, Soro G, Canziani A, Menicanti L, Frigiola A (2004) Reduced systemic heparin dose with phosphorylcholine coated closed circuit in coronary operations. Int J Artif Organs 27(4): 311–319.

Reuben BG, Perl O, Morgan NL, Stratford P, Dudley LY, Hawes C (1995) Phospholipid coatings for the prevention of membrane fouling. J Chem Technol Biotechnol 63(1): 85–91.

Rose SF, Lewis AL, Hanlon GW, Lloyd AW (2004) Biological responses to cationically charged phosphorylcholine-based materials in vitro. Biomaterials 25(21): 5125–5135.

Rose SF, Okere S, Hanlon GW, Lloyd AW, Lewis AL (2005) Bacterial adhesion to phosphorylcholine-based polymers with varying cationic charge and the effect of heparin pre-adsorption. J Mater Sci Mater Med 16(11): 1003–1015.

Ruiz L, Hilborn JG, Léonard D, Mathieu HJ (1998) Synthesis, structure and surface dynamics of phosphorycholine functional biomimicking polymers. Biomaterials 19(11-12): 987–998.

Russell J (2000) Bacteria, biofilms and devices: the possible protective role of phosphorylcholine materials. J Endourol 14(1): 39–42.

Russell JC, Jones JR, Vick TA, Stratford PW (1993) The preparation of tritium labelled biocompatible polymers. J Labelled Comp Radiopharm 33: 957–964.

Sakai-Kato K, Kato M, Ishihara K, Toyo'oka T (2004) An enzyme-immobilization method for integration of biofunctions on a microchip using a water-soluble amphiphilic phospholipid polymer having a reacting group. Lab Chip 4(1): 4–6.

Salvage JP, Rose SF, Phillips GJ, Hanlon GW, Lloyd AW, Ma IY, Armes SP, Billingham NC, Lewis AL (2005) Novel biocompatible phosphorylcholine-based self-assembled nanoparticles for drug delivery. J Control Release 104(2): 259–270.

Sawada S, Iwasaki Y, Nakabayashi N, Ishihara K (2006) Stress response of adherent cells on a polymer blend surface composed of a segmented polyurethane and MPC copolymers. J Biomed Mater Res A 79(3): 476–484.

Seo JH, Matsuno R, Konno T, Takai M, Ishihara K (2008) Surface tethering of phosphorylcholine groups onto poly(dimethylsiloxane) through swelling: deswelling methods with phospholipids moiety containing ABA-type block copolymers. Biomaterials 29(10): 1367–1376.

Shigeta M, Tanaka T, Koike N, Yamakawa N, Usui M (2006) Suppression of fibroblast and bacterial adhesion by MPC coating on acrylic intraocular lenses. J Cataract Refract Surg 32(5): 859–866.

Shin EK, Han SH, et al. (2005) Favourable outcome of high dose dexamethasone eluting biodivysio drug delivery phosphoylcholine coated stent compared with unloaded stent. Eur Heart J 26(1): 706.

Sibarani J, Takai M, Ishihara K (2007a) Surface modification on microfluidic devices with 2-methacryloyloxyethyl phosphorylcholine polymers for reducing unfavorable protein adsorption. Colloids Surf B Biointerfaces 54(1): 88–93.

Sibarani J, Konno T, Takai M, Ishihara K (2007b) Surface modification by grafting with biocompatible 2-methacryloyloxyethyl phosphorylcholine for microfluidic devices. Key Eng Mat 342-343: 789–792.

Snyder TA, Tsukui H, Kihara S, Akimoto T, Litwak KN, Kameneva MV, Yamazaki K, Wagner WR (2007) Preclinical biocompatibility assessment of the EVAHEART ventricular assist device: coating comparison and platelet activation. J Biomed Mater Res A 81(1): 85–92.

Stenzel MH, Barner-Kowollik C, Davis TP, Dalton HM (2004) Amphiphilic block copolymers based on poly(2-acryloyloxyethyl phosphorylcholine) prepared via RAFT polymerisation as biocompatible nanocontainers. Macromol Biosci 4(4): 445–453.

Sumida E, Iwasaki Y, Akiyoshi K, Kasugai S (2006) Platelet separation from whole blood in an aqueous two-phase system with water-soluble polymers. J Pharmacol Sci 101(1): 91–97.

Tang Y, Lu JR, Lewis AL, Vick TA, Stratford PW (2001) Swelling of zwitterionic polymer films characterized by spectroscopic ellipsometry. Macromolecules 34(25): 8768–8776.

Tang Y, Lu JR, Lewis AL, Vick TA, Stratford PW (2002) Structural effects on swelling of thin phosphorylcholine polymer films. Macromolecules 35: 3955–3964.

Tang Y, Su TJ, Armstrong J, Lu JR, Lewis AL, Vick TA, Stratford PW, Heenan RK, Penfold J (2003) Interfacial structure of phosphorylcholine incorporated biocompatible polymer films. Macromolecules 36: 8440–8448.

Ueda H, Watanabe J, Konno T, Takai M, Saito A, Ishihara K (2006) Asymmetrically functional surface properties on biocompatible phospholipid polymer membrane for bioartificial kidney. J Biomed Mater Res A 77(1): 19–27.

Ueda T, Watanabe A, Ishihara K, Nakabayashi N (1991) Protein adsorption on biomedical polymers with a phosphorylcholine moiety adsorbed with phospholipid. J Biomater Sci Polym Ed 3(2): 185–194.

Wada M, Jinno H, Ueda M, Ikeda T, Kitajima M, Konno T, Watanabe J, Ishihara K (2007) Efficacy of an MPC-BMA co-polymer as a nanotransporter for paclitaxel. Anticancer Res 27(3B): 1431–1435.

Walter DH, Cejna M, Diaz-Sandoval L, Willis S, Kirkwood L, Stratford PW, Tietz AB, Kirchmair R, Silver M, Curry C, Wecker A, Yoon YS, Heidenreich R, Hanley A, Kearney M, Tio FO, Kuenzler P, Isner JM, Losordo DW (2004) Local gene transfer of phVEGF-2 plasmid by gene-eluting stents: an alternative strategy for inhibition of restenosis. Circulation 110(1): 36–45.

Wang H, Miyamoto A, Hirano T, Seno M, Sato T (2004) Radical polymerisation of 2-methacryloyloxyethyl phosphorylcholine in water: kinetics and salt effects. Eur Polym J 40(10): 2287–2290.

Watanabe J, Ishihara K (2008) Establishing ultimate biointerfaces covered with phosphorylcholine groups. Colloids Surf B Biointerfaces 65(2): 155–165.

West SL, Salvage JP, Lobb EJ, Armes SP, Billingham NC, Lewis AL, Hanlon GW, Lloyd AW (2004) The biocompatibility of crosslinkable copolymer coatings containing sulfobetaines and phosphobetaines. Biomaterials 25(7-8): 1195–1204.

Whelan DM, van der Giessen WJ, Krabbendam SC, van Vliet EA, Verdouw PD, Serruys PW, van Beusekom HM (2000) Biocompatibility of phosphorylcholine coated stents in normal porcine coronary arteries. Heart 83(3): 338–345.

Xu JP, Ji J, Chen WD, Shen JC (2005a) Novel biomimetic polymersomes as polymer therapeutics for drug delivery. J Control Release 107(3): 502–512.

Xu JP, Ji J, Chen WD, Shen JC (2005b) Novel biomimetic surfactant: synthesis and micellar characteristics. Macromol Biosci 5(2): 164–171.

Xu Y, Takai M, Konno T, Ishihara K (2007a) Coating of anionic phospholipid polymer onto silica-based microchannels to minimise nonspecific protein adsorption. Key Eng Mat 342-343: 797–800.

Xu Y, Takai M, Konno T, Ishihara K (2007b) Microfluidic flow control on charged phospholipid polymer interface. Lab Chip 7(2): 199–206.

Yang S, Zhang SP, Winnik FM, Mwale F, Gong YK (2008) Group reorientation and migration of amphiphilic polymer bearing phosphorylcholine functionalities on surface of cellular membrane mimicking coating. J Biomed Mater Res A 84(3): 837–841.

Yang Y, Zhang SF, Kingston MA, Jones G, Wright G, Spencer SA (2000) Glucose sensor with improved haemocompatibilty. Biosens Bioelectron 15(5-6): 221–227.

Yang Y, Zhang S, Jones G, Morgan N, El Haj AJ (2004) Phosphorylcholine-containing polymers for use in cell encapsulation. Artif Cells Blood Substit Immobil Biotechnol 32(1): 91–104.

Yao K, Huang XD, Huang XJ, Xu ZK (2006) Improvement of the surface biocompatibility of silicone intraocular lens by the plasma-induced tethering of phospholipid moieties. J Biomed Mater Res A 78(4): 684–692.

Ye SH, Watanabe J, Iwasaki Y, Ishihara K (2003) Antifouling blood purification membrane composed of cellulose acetate and phospholipid polymer. Biomaterials 24(23): 4143–4152.

Ye SH, Watanabe J, Ishihara K (2004) Cellulose acetate hollow fiber membranes blended with phospholipid polymer and their performance for hemopurification. J Biomater Sci Polym Ed 15(8): 981–1001.

Ye SH, Watanabe J, Takai M, Iwasaki Y, Ishihara K (2005) Design of functional hollow fiber membranes modified with phospholipid polymers for application in total hemopurification system. Biomaterials 26(24): 5032–5041.

Yokoyama R, Suzuki S, Shirai K, Yamauchi T, Tsubokawa N, Tsuchimochi M (2006) Preparation and properties of biocompatible polymer-grafted silica nanoparticle. Eur Polym J 42(12): 3221–3229.

Yoneyama T, Ito M, Sugihara K, Ishihara K, Nakabayashi N (2000) Small diameter vascular prosthesis with a nonthrombogenic phospholipid polymer surface: preliminary study of a new concept for functioning in the absence of pseudo- or neointima formation. Artif Organs 24(1): 23–28.

Young G, Bowers R, Hall B, Port M (1997a) Clinical comparison of Omafilcon A with four control materials. CLAO J 23(4): 249–258.

Young G, Bowers R, Hall B, Port M (1997b) Six month clinical evaluation of a biomimetic hydrogel contact lens. CLAO J 23(4): 226–236.

Yuan JJ, Armes SP, Takabayashi Y, Prassides K, Leite CA, Galembeck F, Lewis AL (2006a) Synthesis of biocompatible poly[2-(methacryloyloxy)ethyl phosphorylcholine]-coated magnetite nanoparticles. Langmuir 22(26): 10989–10993.

Yuan JJ, Schmid A, Armes SP, Lewis AL (2006b) Facile synthesis of highly biocompatible poly(2-(methacryloyloxy)ethyl phosphorylcholine)-coated gold nanoparticles in aqueous solution. Langmuir 22(26): 11022–11027.

Yusa S, Fukuda K, Yamamoto T, Ishihara K, Morishima Y (2005) Synthesis of well-defined amphiphilic block copolymers having phospholipid polymer sequences as a novel biocompatible polymer micelle reagent. Biomacromolecules 6(2): 663–670.

Zhang S, Benmakroha Y, Rolfe P, Tanaka S, Ishihara K (1996) Development of a haemocompatible pO2 sensor with phospholipid-based copolymer membrane. Biosens Bioelectron 11(10): 1019–1029.

Zhang T, Song Z, Chen H, Yu X, Jiang Z (2008a) Synthesis and characterization of phosphorylcholine-capped poly(epsilon-caprolactone)-poly(ethylene oxide) di-block co-polymers and its surface modification on polyurethanes. J Biomater Sci Polym Ed 19(4): 509–524.

Zhang Z, Cao X, Zhao X, Holt CM, Lewis AL, Lu JR (2008b) Controlled delivery of anti-sense oligodeoxynucleotide from multilayered biocompatible phosphorylcholine polymer films. J Control Release 130(1): 69–76.

Zhao X, Zhang Z, Pan F, Ma Y, Armes SP, Lewis AL, Lu JR (2005) Solution pH-regulated interfacial adsorption of diblock phosphorylcholine copolymers. Langmuir 21(21): 9597–9603.

Zhao X, Pan F, Zhang Z, Grant C, Ma Y, Armes SP, Tang Y, Lewis AL, Waigh T, Lu JR (2007) Nanostructure of polyplexes formed between cationic diblock copolymer and antisense oligodeoxynucleotide and its influence on cell transfection efficiency. Biomacromolecules 8(11): 3493–3502.

Zhao X, Zhang Z, Pan F, Waigh TA, Lu JR (2008) Plasmid DNA complexation with phosphorylcholine diblock copolymers and its effect on cell transfection. Langmuir 24(13): 6881–6888.

Zwaal RF, Comfurius P, van Deenen LL (1977) Membrane asymmetry and blood coagulation. Nature 268(5618): 358–360.

5

BIOMIMETIC, BIORESPONSIVE, AND BIOACTIVE MATERIALS: INTEGRATING MATERIALS WITH TISSUE

Roberto Chiesa and Alberto Cigada

5.1 INTRODUCTION

The unique structural, mechanical, and chemical properties of various metals make them well suited for manufacturing a wide variety of implantable biomedical devices. In particular, their high strength and stiffness are critical factors where metals are used in load-bearing medical applications, such as where the structural function of the musculoskeletal system must be restored. As a consequence, orthopedic and dental applications that require the use of metals [Smith 1993; Katti 2004; Mudgal & Jupiter 2006] commonly opt to use titanium and titanium alloys, cobalt chromium alloys, and stainless steels [Long & Rack 1998; Disegi & Eschbach 2000; Marti 2000; Pohler 2000]. In general, where devices are required to bear high levels of stress with a controlled low strain, the mandatory dimensional and geometrical requirements dictate that metal-based products offer the only possible solution. In addition, metals provide numerous other biomedical functions [Shabalovskaya 1996; Leppaniemi et al. 2000; Yim et al. 2000].

The application of metals in biomedical fields is well established, with a number of well-known metal-based materials used routinely in different situations. Biomedical research involving metals continues to be a very active area, with much current research devoted to modifying and improving metal surfaces to inhibit or promote specific tissue reactions. It is likely that a new generation of biomedical devices will result from the

Biomimetic, Bioresponsive, and Bioactive Materials: An Introduction to Integrating Materials with Tissues, First Edition. Edited by Matteo Santin and Gary Phillips.

current advances in understanding the complex phenomena that occur at the interface between the metal biomaterial surface and biological tissues. The ultimate goal is to improve the functional capability of implanted devices over their lifetime [Pilliar 1991, 1998; Wagner 1992; Dearnaley 1993; Morra & Cassinelli 2006].

In this chapter, the general bulk properties of metals are discussed, with particular reference to selected structural biomedical applications. In addition, some of the current strategies for modifying metal surfaces to optimize implant–tissue integration are discussed.

5.2 MANDATORY REQUIREMENTS FOR METALS AS IMPLANTABLE MATERIALS

Metals must fulfill several mechanical requirements to be used successfully in biomedical applications. These include strength, stiffness, toughness, fatigue resistance, chemical resistance to general and localized corrosion, and biological biocompatibility [Black 1988; Friedman et al. 1993].

General Requirements

In general, any biomaterial used for a structural application, such as in the manufacture of orthopedic prostheses, osteosynthetic devices, or dental implants, must possess

- high mechanical strength, stiffness, and fatigue resistance,
- high resistance to degradation,
- high biocompatibility.

These conditions are so restrictive that they are found among only some metals:

- austenitic stainless steels
- cobalt chromium alloys
- titanium and titanium alloys

5.2.1 Stiffness

Stiffness can be defined as resistance to deformation or deflection of an elastic body, and is one of the most important design factors driving the selection of materials in most structural applications. Stiffness is not only related to the dimension and shape of a body, but is also an intrinsic property of the material known as Young's modulus or elastic modulus (symbol: E). When considering structural applications of biomaterials in orthopedics, the shape and dimensions of hip joint stems and osteosynthesis devices are often predetermined and cannot be modified. To ensure that the device has the appropriate high degree of stiffness, materials with a high elastic modulus must be selected. In the case of orthopedic and dental implants, the elastic modulus must be much higher than that of the bone supported or replaced by the implant. Only the elastic modulus of metals is sufficiently high, typically in the order of hundreds of gigapascals (GPa), to ensure that any load transfer is sufficiently controlled to produce low deformation. Polymeric biomaterials have elastic moduli of the order of only a few GPa (UHMWPE: 0.9–0.96 GPa, PMMA: 2–3 GPa), and this allows excessive deformation of the material with deflection of any loaded device. Hence, they are not compatible with the functional and structural requirements of orthopedic prostheses. The elastic modulus can also vary widely among metallic biomaterials; E is around 110 GPa for titanium and titanium alloys, about 210 GPa for stainless steels, and around 230 GPa for cobalt chromium alloys. The selection of the best metal alloy is often determined by this mechanical parameter.

5.2.2 Strength

The compressive, tensile, and shear strengths of biomaterials also affect their suitability. The tensile strengths can be determined from two parameters: ultimate tensile strength (UTS) and tensile yield strength (TYS), both expressed in megapascals (MPa). A high tensile strength is an essential requirement of any biomaterial for use in musculoskeletal applications. For example, an osteosynthesis device that has a section 10 times less than the bone it substitutes requires strength at least 10 times that of the bone. Consequently, if cortical bone strength is around 100 MPa, any osteosynthetic material must possess a UTS around 1000 MPa. Only a few metal alloys have such high values. A second important consideration is fatigue resistance. Living tissues are capable of regeneration and are able to recover from microfractures, cracking, and other damage. Synthetic biomaterials cannot regenerate, and failure can occur when cyclic stress is applied even at low stress levels as a result of fatigue failure. The long-term performance of a device can be impaired by variations and defects introduced during production, such as inclusions and microdefects, during surgery when handling of the implant scratches the surface, or from degradation due to general or localized corrosion. To maximize the resistance of a biomedical device to fatigue over its lifetime, especially when cyclic stresses are involved, a high-strength material with a high level of fatigue resistance is essential [Krygier et al. 1994]. For example, the materials used in hip and knee replacements must have strength in the range of 700–900 MPa. Such stringent mechanical requirements limit the choice of biomaterials for such applications. Manufacturing technologies impose further major limitations on the choice of material. For example, casting may provide the most suitable technology to manufacture the stems of hip prostheses or of femoral components or knee prostheses with a complex shape. In such cases, only a particular cobalt-based alloy may combine the required high mechanical strength with the properties that allow casting to be used.

As a result of these important restrictive mechanical considerations, no polymeric or ceramic materials available at present can be used, and only a few titanium alloys, stainless steels, and cobalt chromium alloys are available for such applications. Although composite materials may possess the necessary high strength and high stiffness to satisfy the mechanical demands, their clinical application is nowadays limited by biocompatibility issues.

5.2.3 Corrosion Resistance

Only few inert metals, such as gold and platinum, are thermodynamically immune from corrosion in the human body. Unfortunately, the mechanical properties of these metals are unsatisfactory for most biomedical applications, and their use is limited to specific applications.

None of the high-strength metals are immune from corrosion in body fluids. In normal situations, their degradation and dissolution rates are very low, and they are generally well tolerated by the biological system. Nevertheless, a high rate of ion release can be induced by corrosion. In such cases, the chemical compositions of the metal alloys play a major role in their biocompatibility, and local or systemic reactions to some metal ions may impair their integration with tissues.

5.2.3.1 General Corrosion. Metals used for biomedical applications must possess a high degree of corrosion resistance. Biological fluids that are rich in chloride and oxygen are extremely corrosive to most metals. Corrosion in body fluid always involves two simultaneous electrochemical half reactions: the oxidation of the metal that causes corrosion and the release of metals ions, and the complementary reduction of oxygen. For example, the corrosion of iron in an oxygen-rich solution can be described by

$$2Fe \rightarrow 2Fe^{2+} + 4e^- \text{ (anodic reaction: oxidation half reaction)}$$

$$O_2 + 2H_2O + 4e^- \rightarrow 4OH^- \text{ (cathodic reaction: reduction half reaction)}$$

Every oxidation and reduction half reaction is characterized by a thermodynamic potential; the higher the potential, the lower the susceptibility of the metal to corrosion. If the oxidation potential of the metal is higher than the potential of the reduction half reaction, the metal is thermodynamically immune from corrosion in the environment in question. Corrosion can occur if the thermodynamic oxidation potential is lower than the potential of the complementary reduction half reaction.

Only a few inert metals, such as gold and platinum, possess sufficiently high oxidation potentials to render them immune from corrosion in the human body. All other metals have lower oxidation potential and can be corroded.

Nevertheless, some special metals, such as stainless steels and titanium and cobalt alloys, are characterized by high kinetic dissipation as a result the formation of a protective surface oxide layer known as the passivation film. Even if such alloys are thermodynamically susceptible to corrosion, their corrosion rate is extremely low and can be considered negligible for many applications. Under normal conditions, the ion release rate of such passivated metals resulting from general corrosion is usually well tolerated by biological systems. Nevertheless, some specific conditions may promote corrosion and increase the ion release rate by several orders of magnitude, such as crevice corrosion that can occur in stainless steels or fretting corrosion associated with titanium alloys.

5.2.3.2   Crevice Corrosion. Crevice corrosion can affect stainless steel components such as osteosynthesis devices [Rondelli et al. 1997]. The process is related to the penetration of fluids into crevices and poorly shielded spaces where body fluid can stagnate. Typical locations in osteosynthesis devices include screw heads and holes in metal plates. Any modification of the fluid composition through aeration or changes in pH can increase the local corrosion potential and result in high levels of corrosion and elevated ion release rates in the surrounding area. Crevice corrosion is particularly common in stainless steels [Mansfeld et al. 1992], with cobalt chromium and titanium alloys proving far less susceptible. This is one of the most important reasons for the initiation of fatigue failure in mechanically stressed stainless steel biomedical components and can result in a 100-fold increase in metal ion release.

5.2.3.3   Fretting Corrosion. Rubbing movements with small amplitudes of no more than a few micrometers on a passivated metal surface can lead to mechanical damage of the protective oxide film (passivated layer) covering the surface. This fretting can lead to an increase in the otherwise very low dissolution rate of metal ions. Titanium and titanium alloys are particular prone to this type of corrosion, which can initiate fatigue cracks resulting in the subsequent failure of titanium-based structural components.

5.2.3.4   Galvanic Corrosion. Galvanic corrosion can occur when a metal is electrochemically linked to another metal possessing a higher thermodynamic potential (i.e., more inert). Such corrosion is now rare and is usually avoided by the appropriate selection of different metal components. Of the three main classes of metal used for biomaterials, the juxtaposition of either titanium or cobalt chromium with stainless steel should be avoided, whereas coupling of titanium with cobalt chromium does not present any particular problem.

5.3   BIOCOMPATIBILITY OF METALS

As outlined above, when titanium, cobalt chromium, or stainless steels are used as structural metals in biomedicine, they always release some ions into the body fluids and the surrounding tissues. The chemical compositions of the metal alloys, their release rates, and the cumulative amounts can have important implications for the general biocompatibility of the implanted medical device, with all three factors having the potential to impact the biological system. Although the slow release of ions resulting from general corrosion cannot be avoided completely, the localized corrosion that is responsible for increased levels of ion dissolution must be limited and controlled. Some metal ions, such as iron, are tolerated well and easily eliminated by the body through normal physiological processes. Other metal ions, for example, chrome, nickel, and cobalt, accumulate in specific organs such as the spleen, liver, and kidney [Pascual et al. 1992]. Nickel biocompatibility is of particular concern in the biomedical field [Shabalovskaya 2002]. There is increasing concern regarding the problems associated with contact dermatitis and allergic reactions induced by skin contact with nickel.

Similar problems are also experienced with implanted metal components that contain this element, with all conventional implanted stainless steels and cobalt chromium alloys containing nickel. For mechanical reasons, it is very difficult to make alloys of high strength and high stiffness that lack nickel. To circumvent this difficulty, companies have been evaluating a range of surface treatments, such as use of ceramic coatings, with the aim of sealing off metal surfaces to reduce corrosion and hence control metal ion release into biological tissues. Some new alloy formulations with very low nickel contents are also under consideration, and implantable devices developed in the past decade using these low-nickel alloys have been shown to be especially resistant to localized corrosion. The use of high nitrogen stainless steels that comply with ISO standard 5832-9 "Implants for surgery—Metallic materials—Part 9: Wrought high nitrogen stainless steel," and the development of further new alloys with very low nickel contents [Fini et al. 2003] may help to resolve this important biocompatibility issue.

In addition to the chemical compositions and nickel contents of metal alloys, the release rates of the metal ions and their localized corrosion resistance are also key factors in improving the biocompatibility of nickel-bearing metal alloys. The control of localized corrosion can be achieved through the appropriate choice of material and proper design of the device to ensure there is an absence of closed areas where fluids can stagnate and initiate crevice corrosion.

5.3.1 ISO Standardized Metal Family

Most of the metals used for biomedical applications are listed in the ISO 5832 standard. They include the three families of metals discussed above: stainless steels, cobalt chromium alloys, and titanium alloys. Manufacturers of biomedical devices can refer to the standard to select the appropriate material for a given application. As long as the producer chooses one of the materials listed in ISO 5832, the certification procedure necessary for safe commercial development of the product, such as gaining U.S. Food and Drug Administration (FDA) approval, is greatly simplified given that the material has already been fully tested for biomedical applications. The different metals described in ISO 5832 possess very different mechanical properties and corrosion responses, and selection of the appropriate metal for a specific application requires consideration of a variety of parameters.

5.3.1.1 Stainless Steels. Stainless steels are widely used in orthopedic applications, such as temporary implantable (osteosynthesis) devices. The main advantages of these materials are related to their good mechanical properties obtained after strain hardening, their low cost, and good workability either via plastic deformation or tool machining. Negative aspects of stainless steels include the presence of nickel in the alloys and the possibility of crevice corrosion in some alloy formulations.

ISO standard 5832 (Table 5.1) provides a description of three classes of stainless steel:

- ISO 5832-1 Composition D
- ISO 5832-1 Composition E
- ISO 5832-9

TABLE 5.1. Chemical Composition and Mechanical Properties in the Annealed and in the Work-Hardened States of Different ISO Standardized Metallic Materials Used for Biomedical Applications

Class	Kind	ISO Classification	Chemical Composition (%)	UTS (MPa) Annealed—Cold Worked	TYS (MPa) Annealed—Cold Worked
Austenitic stainless steels	AISI 316L	ISO 5832-1 D	Fe = balance, Cr = 17–19, Ni = 13–15, Mo = 2.25–3.5, N < 0.10	690–1100	190–690
	AISI 317L	ISO 5832-1 E	Fe = balance, Cr = 17–19, Ni = 14–16, Mo = 2.35–4.2, N = 0.1–0.2	800–1100	285–690
	High N	ISO 5832-9	Fe = balance, Cr = 19.5–22, Ni = 9–11, Mo = 2–3, Mn = 2–4.25, N = 0.25–0.5	740–1800	430–n.d.
Cobalt alloys	Cast	ISO 5832-4	Co = balance, Cr = 26.5–30, Mo = 4.5–7	665	450
	Wrought	ISO 5832-5	Co = balance, Cr = 19–21, W = 14–16, Ni = 9–11	860	310
		ISO 5832-6	Co = balance, Ni = 33–37, Cr = 19–21, Mo = 9–10.5	800–1200	300–1000
		ISO 5832-7	Co = 39–42, Cr = 18.5–21.5, Ni = 15–18, Mo = 6.5–7.5, Fe = balance	950–1450	450–1300
		ISO 5832-8	Co = balance, Ni = 15–25, Cr = 18–22, Mo = 3–4, W = 3–4, Fe = 4–6	600–1580	275–1310
Titanium and titanium alloys	Titanium	ISO 5832-2 G1	Ti = balance, O < 0.18	240	170
		ISO 5832-2 G2	Ti = balance, O < 0.25	345	230
		ISO 5832-2 G3	Ti = balance, O < 0.35	450	300
		ISO 5832-2 G4	Ti = balance, O < 0.45	550–680	440–520
	Ti6Al4V	ISO 5832-3	Ti = balance, Al = 5.5–6.75, V = 3.5–4.5	860	780
	Ti5Al2,5Fe	ISO 5832-10	Ti = balance, Al = 4.5–5.5, Fe = 2.5–3	900	800
	Ti7Al8Nb	ISO 5832-11	Ti = balance, Al = 5.5–6.75, Nb = 6.5–7.5	900	800

147

ISO 5832-1 Composition D is a traditional stainless steel, roughly corresponding to the well-known AISI 316L steel. It contains chromium (17–19%), nickel (13–15%), molybdenum (2.25–3.5%), and nitrogen (less than 0.1%). In comparison to AISI 316L, ISO 5832-1 Composition D is characterized by higher purity with lower sulfur, phosphorus, and inclusion contents. This is the cheapest metallic material listed in the ISO 5832 standard and can be easily worked either by plastic deformation or tool machining. Its mechanical properties are generally low in an annealed state (UTS = 690 MPa, TYS = 190 MPa) but can be increased significantly through cold plastic deformation (UTS = 1100 MPa, TYS = 690 MPa). Nevertheless, this type of stainless steel is susceptible to crevice corrosion, and its use should be limited.

ISO 5832-1 Composition E stainless steel contains chromium (17–19%), nickel (14–16%), molybdenum (2.35–4.2%), and nitrogen (0.1–0.2%). With its higher molybdenum and nitrogen contents as compared to ISO 5832-1 Composition D, Composition E shows greater resistance to crevice corrosion but is still not completely immune from this type of corrosion.

ISO 5832-9 stainless steel contains chromium (19.5–22%), nickel (9–11%), molybdenum (2–3%), manganese (2–4.25%), and nitrogen (0.25–0.5%); the high nitrogen level ensures a high degree of resistance to crevice corrosion and good mechanical properties in either the annealed (UTS = 740 MPa, TYS = 430 MPa) or cold-worked condition (TYS up to 1800 MPa). However, the higher cost as compared to ISO 5832-1 Compositions D and E and the difficulty of working by either tool machining or plastic deformation place limits on its use despite its superior properties.

New formulations of stainless steels that have proven highly resistant to corrosion and that have very low nickel contents have been developed. Commercial names include Boheler P558, Carpenter BioDur 108, Krupp Macrofer 2515 MoN. Such innovative stainless steels, which comply with the mechanical requirements of orthopedic applications and are highly resistant to localized corrosion in the biological fluids, have opened some interesting perspectives for producing nickel-free metallic implantable devices with high strength and high elastic modulus.

5.3.1.2 Cobalt Alloys

CAST COBALT ALLOYS. Cast cobalt alloys have long been used in the biomedical field. They possess outstanding mechanical properties and have a high corrosion resistance and good biocompatibility. The strength and fatigue resistance of such alloys is strongly related to the metallurgical quality of the castings.

The main advantages of using these alloys are that they allow casting technology to be used to produce implants with high strength and high elastic modulus, and that have good corrosion resistance, especially to fretting corrosion, coupled with excellent biocompatibility. The main disadvantages are related to the high cost of the alloy, the low fatigue resistance that is dependent on the presence of metallurgical defects, the inability to employ plastic deformation, and difficulties with tool machining.

Cast cobalt alloys (ISO 5832-4) contain chromium (26.5–30%) and molybdenum (4.5–7%). The UTS reaches 650 MPa, while TYS can attain 450 MPa.

WROUGHT COBALT CHROMIUM ALLOYS. Wrought cobalt chromium alloys can be processed by plastic deformation technology. Such alloys possess good mechanical properties and good corrosion resistance. Their high nickel content, the complexity of the production technology, and the high cost of the alloy are the main disadvantages of such alloys. The ISO standard currently lists five formulations for wrought cobalt alloys (Table 5.1): ISO 5832-5, ISO 5832-6, ISO 5832-7, ISO 5832-8, and ISO 5832-12.

5.3.1.3 Titanium and Titanium Alloys.

Titanium possesses the best biocompatibility among high-strength implantable metals. If high resistance is the primary requirement, then titanium alloys must be used. Compared to pure titanium, its alloys are more expensive and show lower biocompatibility, but have much higher mechanical properties. However, titanium and its alloys have an elastic modulus of around 110 GPa, half those of stainless steels and cobalt alloys. Such a low modulus in comparison to other metals makes titanium of particular interest in applications where higher deflection and deformation are advantageous. For example, titanium alloys can be used for manufacturing noncemented hip stems, to decrease the stiffness of the stem, and to provide load transfer in the proximal region of the femur, therefore decreasing stress shielding problems [Head et al. 1995].

Titanium and titanium alloys show excellent resistance to crevice corrosion but poor resistance to fretting corrosion. Such localized corrosion can be important where different components are connected together, such as in screw/plate connections in osteosynthesis devices, in taper/head connections, and in modular hip prosthesis designs. Low fretting corrosion resistance is still one of the main limitations of the use of titanium alloys in biomedicine.

PURE TITANIUM. Pure titanium shows very good biocompatibility due to its osseointegration capability, and it has good workability by either tool machining or plastic deformation. However, its poor mechanical properties and its sensitivity to fretting corrosion pose limitations in its use.

The ISO 5832-2 standard lists four types of commercially pure titanium (grades 1, 2, 3, 4) with oxygen contents varying from 0.18% to 0.40%. The increase in oxygen corresponds to an increase in the mechanical properties, with UTS varying from 240 MPa for grade 1 to 550 MPa for annealed grade 4 and 680 MPa for cold-worked grade 4. This increase in strength is always accompanied by a decrease in elongation and an increase in difficulties in working the alloy using tool machining. The mechanical properties of pure titanium are often well suited to the requirements of dental implants, with commercially pure grade 2 and grade 4 titanium often used in this field. Most orthopedic prosthetic applications usually require much higher strength that pure titanium can provide and require the use of titanium alloys.

TITANIUM ALLOYS. The ISO 5832 standard lists three types of titanium alloy: Ti6Al4V (ISO 5832-3), Ti5Al2.5Fe (ISO 5832-10), and Ti6Al7Nb (ISO 5832-11). The most commonly used titanium alloy is ISO 5832-3 with a UTS of 860 MPa in the annealed state. ISO 5832-10 and ISO 5832-11 alloys are very similar to Ti6Al4V in terms of corrosion resistance and mechanical properties but are more expensive and as a consequence are used less frequently.

5.4 SURFACE TREATMENTS OF METALS FOR BIOMEDICAL APPLICATIONS

Although the long-term performance of metallic implantable materials is principally related to the mechanical properties of the bulk metal and to the biomechanical performance of the device, modification of the surface properties offers the possibility of exerting some control over those processes relevant to biological tissue integration.

A wide variety of surface treatments have been applied to metal components to improve specific properties. In general, every metal device is subject to an appropriate surface finishing process at the end of its manufacture. Many traditional industrial finishes, including sandblasting, pickling, etching, degreasing, and polishing, have been commonly applied to metallic implants. Other surface treatments that are specific to the biomedical field include the deposition of metals, ceramics, and polymers employing a variety of coating procedures. Primarily, surface treatments are applied to improve the mechanical performance or biological compatibility, or to trigger specific biological responses. For example, the surface roughness of noncemented hip prosthetic stems is commonly increased by a mechanical or physical finishing process, such as grit blasting, deposition of a porous metal film, or coating with metal microfilaments or microbeads. The increase in roughness ensures better load transfer. It provides a wider contact area and ensures stronger mechanical interconnection with bone tissue that grows on the metal interface [Browne & Gregson 1994; Piattelli et al. 1996; Hanawa 1999]. Elsewhere, the wear resistance of a hip or knee metal joint can be improved by polishing and hardening the metal surface. Hard ceramic coatings can help to reduce metal ion release and improve wear performance of sliding bearing surfaces [Jones et al. 2001]. Metal osteosynthesis devices are usually finished by mechanical or electrochemical polishing to increase their fatigue resistance and to reduce localized corrosion potential [Pourbaix 1984].

Although most finishing techniques applied to metallic biomedical devices were not developed specifically with biomedical applications in mind, a considerable proportion of current biomaterials research is focused on developing specific new treatments capable of modifying surface properties to address precise biological needs. Titanium and its alloys are of particular interest due to their superior mechanical properties, optimal biocompatibility, and especially to the intriguing if somewhat peculiar properties of their surface oxide layers. These surface layers can be used to modify the interface properties of implants [Southerland et al. 1993; Ong et al. 1995]. They are composed mainly of TiO_2 and grow spontaneously on titanium in the presence of oxygen to a thickness of approximately 5 nm [Vezeau et al. 2000]. Thermodynamically, they are very stable. Despite their presence, however, some reactions still occur with surrounding tissue and body fluids [Albreksson et al. 1983; Nanci & Poleo 1999], but an increase in the thickness of this layer may help reduce dissolution of the titanium and titanium alloy elements in body fluids.

Over the last several decades, there have been a number of attempts to improve the osseointegration of titanium. These have included treatments developed to modify the surface morphology and topography of metal implants as well as their chemical composition at the interface. A significant proportion of the noncemented hip prostheses currently on the market in Europe have hydroxyapatite (HA) coatings, usually applied by plasma spray technology [Duheyne & Healy 1988]. Thick HA ceramic coatings, commonly ranging from 40 to 150 μm, improve implant-to-bone bonding and hence reduce the formation of fibrous tissue at the interface [Chae et al. 1992; Tisel et al. 1994]. Nevertheless, some issues remain concerning possible term failures related to adhesion and cohesion of plasma-sprayed HA coatings and their dissolution rates, which are dependent largely on the crystallinity of the HA [Bauer et al. 1991; Buma & Gardeniers 1995]. These questions have led researchers to look for alternative surface modification methods, such as electrodeposition, magnetron sputtering, and ion beam sputtering [Cigada et al. 1993; Fujihiro et al. 1998].

Electrochemical coating techniques are proving of particular interest given their versatility. They allow for cathodic deposition of different calcium phosphate phases and were investigated extensively by many groups in the 1990s [Shirkhanzadeh 1991, 1995; Redepenning et al. 1996]. Deposition using cathodic polarization and subsequent thermal treatments allow titanium and other metals to be coated effectively with calcium phosphate (Figure 5.1). However, such methods allow only suboptimal mechanical adhesion of the film to the substrate in comparison with plasma spray coating technologies. Nonetheless, there has been renewed interest in cathodic polarization due to its

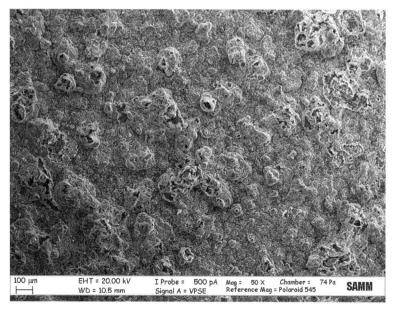

Figure 5.1. SEM image of a calcium phosphate coating deposited by cathodic polarization on titanium substrate.

ability to coat porous structures and to allow doping of the coating with drugs, other macromolecules, and thermally degradable materials.

Anodic oxidation is also suitable for modifying titanium surfaces. This technique makes it possible to predetermine the surface oxide layer thickness, its chemical composition, and its morphology. Anodic spark deposition (ASD) or anodic spark discharge is a high-voltage anodic oxidation technique that allows a thin, porous surface layer to be applied to titanium [Kurze et al. 1984]. By combining ASD with other chemical or physical processes, it is possible to design surfaces of specific chemical compositions and morphologies. The mechanical stability of the anodized layer is very high, and chemical species other than calcium and phosphate can be inserted into the layer.

5.4.1 Cathodic Deposition Treatments

When a titanium cathode is placed in an electrochemical solution, the reactions that take place on its surface depend on the applied voltage, current, and composition and pH of the solution.

Studies of the deposition of bioactive calcium phosphate ceramic coatings formed by cathodic deposition on titanium and titanium alloys performed in the 1990s employed numerous electrolytes [Shirkhanzadeh 1991, 1995]. These included calcium phosphate salts in aqueous solutions of sodium chloride, as well as $Ca(NO_3)_2$ with $NH_4H_2PO_4$ and $CaCl_2$ with $NH_4H_2PO_4$. The properties of the resulting coatings were dependent on the electrolyte composition, including concentration and pH, on the applied voltage, electrolyte temperature, cathode surface extension, and solution agitation. Treatments were performed at relatively low polarization voltages with the coating then subjected to hydrothermal or thermal calcination treatments to induce formation of HA layers. Subsequently, other authors employed cathodic polarization of titanium in a $Ca(H_2PO_4)_2$ electrolyte over a wide potential range to produce a coating of the calcium phosphate phase brushite, which was then transformed into HA in the solid state [Redepenning et al. 1996]. In general, the advantages of cathodic deposition are related to the cost-effectiveness of the technique and the possibility of doping a calcium phosphate coating with ions, such as magnesium and manganese, capable of improving tissue integration by triggering specific biological/biochemical effects. Of particular interest is the possibility offered by the cathodic technique to dope the calcium phosphate layer with metal ions known to possess antibacterial effects, such as silver (Figure 5.2). Such modification can occur during or after electrochemical deposition.

The application of calcium phosphate coatings by cathodic deposition provides an interesting alternative to plasma spray deposition for enhancing the osseointegration capabilities of orthopedic and dental implants, with a series of new industrial products now being proposed for clinical applications [Becker et al. 2004].

5.4.2 Anodic Oxidation

Anodic oxidation electrode reactions coupled with electrical field-driven metal and oxygen ion diffusion can produce an oxide film at the anode surface. Anodization alone can yield an adherent thin film at the metal surface. Titanium colorization is achieved

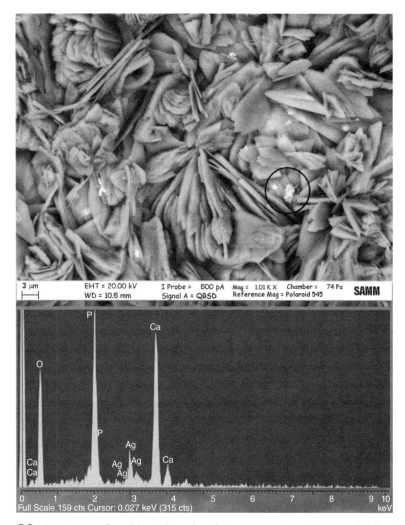

Figure 5.2. SEM image of a calcium/silver phosphate coating deposited by cathodic polarization on titanium substrate. The energy dispersive spectroscopy (EDS) analysis performed on the circle marker evidenced the presence of silver containing crystals.

through low and medium voltage anodization by modifying parameters, such as anode potential, current density, electrolyte composition, and processing temperature.

The anodizing process can be carried out at both constant voltage and constant current. The resulting oxide film exhibits high resistivity, and the oxide will continue to grow as long as the applied electric field is sufficient to drive the ions through the film. The growth of the oxide film during the process is almost linearly dependent on the applied voltage until breakdown of the film occurs between 100 and 160 V depending on the nature of the electrolyte. When anodization is carried out at voltages higher than the dielectric limit of the formed film, the breakdown of the dielectric can occur by sparking, leading to an increase in film thickness and to the development of a porous oxide morphology.

Anodic films obtained below the dielectric limit have been widely studied [Delplancke & Winand 1973; Larsson et al. 1994; Walivaara et al. 1994]. Bright colors can be obtained in many electrolytes and arise from light interference through different oxide thicknesses. Although the surface layer formed by anodization is mainly composed of titanium dioxide, electrolyte elements may also be embedded [Delplancke & Winand 1973; Ask et al. 1988]. Diffraction studies have demonstrated the possibility of modifying the structure of the titanium oxide to that of the tetragonal phase anatase and rutile or orthorhombic brookite [Aladjen 1973]. The anatase structure has been studied extensively due to its photocatalytic and antibacterial properties [Chen et al. 2005; Shieh et al. 2006].

If the dielectric breakdown potential is exceeded during the anodization process, the titanium surface layer will experience spark discharge, and ASD can occur. ASD is an electrochemical technique that induces formation of porous ceramic coatings (Figure 5.3) on a metallic surface during the dielectric breakdown of the insulating layer. ASD involves incorporation of at least one metal ion from the electrolyte solution into the ceramic layer allowing different chemical elements to be inserted conveniently into the growing surface coating. Different aqueous solutions and electrochemical conditions can provide films with different structures, compositions, and morphologies. Several

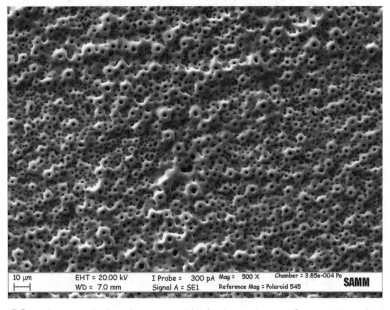

Figure 5.3. Calcium and phosphorous enriched microporous surface coating obtained by anodic spark deposition on titanium substrate.

variables determine the characteristics of the coatings: (1) the composition of the electrolyte, (2) the current density, (3) the voltage, and (4) the temperature and timing of spark discharge.

ASD deposition uses the metal as the positive electrode in the electrochemical circuit. The process in the electrolytic cell consists of three stages: (1) anodization, (2) dielectric breakdown, and (3) coating buildup. Initially, a dielectric layer forms at the surface of the anode and increases during low-voltage anodization. If the current is kept constant, the voltage will grow linearly. During the interval prior to breakdown, most metals produce films displaying a sequence of interference colors.

With a progressive increase in potential, the dielectric film will eventually break down starting with areas corresponding to microscopic film defects, inclusions, morphological discontinuities, and any nonuniform oxide thickness. Where dielectric breakdown occurs, a high current intensity flows through a small area and attracts anions to the metal–coating interface. The temperature increases locally to very high levels and causes fast melting and solidification of material with consequent buildup of the coating. In such spark discharges, the molten spots of material are quenched immediately to the temperature of the electrolyte, and a cascade process takes place that promotes a random spread of sparks over the whole surface. Eventually, the oxide layer increases to cover the entire metal surface. As there is a great deal of interest in its potential uses in biomedicine, the properties and the structures of the films formed by ASD have been studied extensively. Coatings formed by ASD are generally hard and exhibit a porous texture [Wirtz et al. 1991]. It has been demonstrated that the adhesive strength of the ASD film can be optimized using low electrolyte concentrations. The interface between the oxide layer and the metal substrate usually does not show any discontinuity. ASD processing parameters, such as electrolyte temperature, concentration, and flow state, current density, and final voltage, influence the microstructure of the final layer–coating interface [Kurze et al. 1986; Wirtz et al. 1991]. The modified surface film may be crystalline or amorphous depending on the species involved in the melting reaction and on the quenching time of the molten material. The porous morphology of the ASD film is of particular interest for integration of implants with biological tissues. The films often show interconnected porosity with an overlapping structure in which the pore dimensions vary from 200 nm to 2–3 μm, depending on the processing parameters used. In general, the pore diameters increase with increasing current density.

Many different oxide compositions and morphologies have been developed for use in the field of biomedical implants through ASD, including the use of solutions containing calcium phosphate/sodium carbonate/hydrofluoric acid mixtures [Schreckenbach et al. 1999] and solutions rich in calcium and phosphate ions [Yerochin et al. 2000; Sandrini et al. 2005]. Moreover, the possibility of inserting various chemical species into the thin layer, such as potassium, sodium, and silicon (Figure 5.4), offers the potential for providing specific biological solutions for implant–tissue integration. ASD surfaces have also been suggested as matrices to allow incorporation and release of drugs into the surrounding biological environment. This can be achieved by controlling the open porosity of ASD-engineered surfaces [Dunn et al. 1994] or by utilizing specific TiO_2 film morphologies and structures, such as micro- and nanotube conformations.

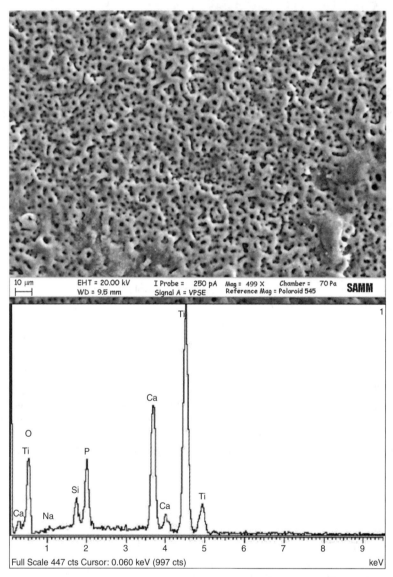

Figure 5.4. Surface morphology of anodic spark deposition treatment performed on titanium. The EDS analysis shows the presence of calcium, phosphorous, silicon, and sodium.

Electrochemical treatment of metals allows modification of the physical structure of the interfacial surface layer and its chemical composition. Moreover, a doped layer of the appropriate morphology can be obtained to promote a specific tissue response, such as osseointegration. Some industrial applications are already available with preliminary clinical treatments providing some encouraging results [Chiesa et al. 2007].

Clearly, electrochemical and, in particular, anodic techniques hold a great deal of promise for the development of a new generation of implantable metal devices.

EXERCISES/QUESTIONS FOR CHAPTER 5

1. List the main physicochemical property requirements of metal-based implants.
2. Provide at least two typical examples of use of metal devices in clinics and discuss their performance and main limitations.
3. What are the main metal biomaterials employed in hip implant manufacturing?
4. What are the main metal biomaterials employed in cardiovascular stent manufacturing?
5. Critically discuss at least two surface treatments of metal biomaterials and their advantage in terms of device biocompatibility and clinical performances.
6. Present a typical surface treatment leading to a biomimetic and/or bioactive approach to tissue repair.
7. Describe two of the main surface deposition methods.
8. What are the mechanisms of metal corrosion upon implantation and their main consequences on host response and implant failure?
9. Present the main advantages offered by the use of alloys in biomedical implants.
10. Describe the anodic spark deposition process and its potential application in biomedical implant manufacturing.

REFERENCES

Aladjen A (1973) Anodic oxidation of titanium and its alloy. J Mater Sci 8: 688–704.

Albreksson T, Hansson H-A, Kasemo B, Larson K, Lundstrom I (1983) The interface zone of inorganic implants in vivo: titanium implants in bone. Ann Biomed Eng 11: 1–27.

Ask M, Lausmaa J, Kasemo B (1988) Preparation and surface spectroscopic characterisation of oxide films on Ti6Al4V. Appl Surf Sci 35: 283–301.

Bauer TW, Geesink RC, Zimmermann R, MaMahon JT (1991) Hydroxyapatite-coated femoral stems: histological analysis of components retrieved at autopsy. J Bone Joint Surg 73: 1439–1452.

Becker P, Neumann HG, Nebe B, Luthen F, Rychly J (2004) Cellular investigations on electrochemically deposited calcium phosphate composites. J Mater Sci Mater Med 15(4): 437–440.

Black J (1988) Orthopedic Biomaterials in Research and Practice. New York: Churchill-Livingstone.

Browne M, Gregson PJ (1994) Surface modification of titanium alloy implants. Biomaterials 15: 894–898.

Buma P, Gardeniers JW (1995) Tissue reaction around a hydroxyapatite-coated hip prosthesis: case report of a retrieved specimen. J Arthroplasty 10: 389–395.

Chae JC, Collier JP, Mayer MB, Surprenant VA, Dauphinais LA (1992) Enhanced ingrowth of porous-coated CoCr implants plasmaspray with tricalcium phospate. J Biomed Mater Res 56: 93–102.

Chen X, Lou Y, Dayal S, Qiu X, Krolicki R, Burda C, Zhao C, Becker J (2005) Doped semiconductor nanomaterials. J Nanosci Nanotechnol 5(9): 1408–1420. Review.

Chiesa R, Giavaresi G, Fini M, Sandrini E, Giordano C, Bianchi A, Giardino R (2007) In vitro and in vivo performance of a novel surface treatment to enhance osseointegration of endosseous implants. Oral Surg Oral Med Oral Pathol Oral Radiol Endod 103(6): 745–756.

Cigada A, De Santis G, Gatti AM, Roos A, Zaffe D (1993) In vivo behavior of a high performance duplex stainless steel. J Appl Biomater 4(1): 39–46.

Dearnaley G (1993) Diamond-like carbon: a potential means of reducing wear in total joint replacements. Clin Mater 12(4): 237–244.

Delplancke JL, Winand R (1973) Galvanostatic anodization of titanium—I. Structures and composition of the anodic films. Electrochim Acta 33(11): 1539–1549.

Disegi JA, Eschbach L (2000) Stainless steel in bone surgery. Injury 31(Suppl. 4): 2–6.

Duheyne P, Healy KE (1988) The effect of plasma-sprayed calcium phosphate ceramic coatings on the metal ion release from porous titanium and cobalt-chromium alloys. J Biomed Mater Res 22: 1137–1163.

Dunn DS, Raghaven S, Volz RG (1994) Ciprofloxacin attachment to porous coated titanium surfaces. J Appl Biomater 5(4): 325–331.

Fini M, Nicoli Aldini N, Torricelli P, Giavaresi G, Corsari V, Lenger H, Bernauer J, Giardino R, Chiesa R, Cigada A (2003) A new austenitic stainless steel with negligible nickel content: an in vitro and in vivo comparative investigation. Biomaterials 24(9): 4929–4939.

Friedman RJ, Black J, Galante JO, Jacobs JJ, Skinner HB (1993) Current concepts in orthopaedic biomaterials and implant fixation. J Bone Joint Surg 75(7): 1086–1109.

Fujihiro Y, Satop N, Uccida S, Sato T (1998) Coating of CaTiO3 on titanium substrates by hydrothermal reactions using calcium-ethylene diamine tetraacetic acid chelate. J Mater Sci Mater Med 9: 363–367.

Hanawa T (1999) In vivo metallic biomaterials and surface modification. Mater Sci Eng A 267: 260–266.

Head WC, Bauk DJ, Emerson RH Jr (1995) Titanium as the material of choice for cementless femoral components in total hip arthroplasty. Clin Orthop Relat Res 311: 85–90. Review.

Jones VC, Barton DC, Auger DD, Hardaker C, Stone MH, Fisher J (2001) Simulation of tibial counterface wear in mobile bearing knees with uncoated and ADLC coated surfaces. Biomed Mater Eng 11(2): 105–115.

Katti KS (2004) Biomaterials in total joint replacement. Colloids Surf B Biointerfaces 39(3): 133–142.

Krygier JJ, Dujovne AR, Bobyn JD (1994) Fatigue behavior of titanium femoral hip prosthesis with proximal sleeve-stem modularity. J Appl Biomater 5(3): 195–201.

Kurze P, Dittrich KH, Krysmann W, Schneider HG (1984) Structure and properties of ANOF layers. Cryst Res Tech 19: 93–99.

Kurze P, Krysmann W, Schneider HG (1986) Applications fields of ANOF layers and composites. Cryst Res Tech 21: 1603–1609.

Larsson C, Thomsen P, Lausmaa J, Rodahl M, Kasemo B, Ericson LE (1994) Bone response to surface modified titanium implants: studies on electropolished implants with different oxide thicknesses and morphology. Biomaterials 15: 1062–1074.

Leppaniemi A, Rich N, Pikoulis E, Rhee P, Burris D, Wherry D (2000) Sutureless vascular reconstruction with titanium clips. Int Angiol 19(1): 69–74.

Long M, Rack HJ (1998) Titanium alloys in total joint replacement—a materials science perspective. Biomaterials 19(18): 1621–1639.

Mansfeld F, Tsai R, Shih H, Little B, Ray R, Wagner P (1992) An electrochemical and surface analytical study of stainless-steels and titanium exposed to natural seawater. Corros Sci 33(3): 445–456.

Marti A (2000) Cobalt-base alloys used in bone surgery. Injury 31(Suppl. 4): 18–21.

Morra M, Cassinelli C (2006) Biomaterials surface characterization and modification. Int J Artif Organs 29(9): 824–833.

Mudgal CS, Jupiter JB (2006) Plate and screw design in fractures of the hand and wrist. Clin Orthop Relat Res 445: 68–80.

Nanci A, Poleo DA (1999) Understanding and controlling the bone–implant interface. Biomaterials 20: 2311–2321.

Ong JL, Lucas LC, Raikar GN, Connatser R, Gregory JC (1995) Spectrospopic characterization of passivated titanium in a physiologic solution. J Mater Sci Mater Med 6: 113–119.

Pascual A, Tsukayama DT, Wicklund BH, Bechtold JE, Merritt K, Peterson PK, Gustilo RB (1992) The effect of stainless steel, cobalt-chromium, titanium alloy, and titanium on the respiratory burst activity of human polymorphonuclear leukocytes. Clin Orthop Relat Res 280: 281–288.

Piattelli A, Scarano A, Piattelli M, Calabrese L (1996) Direct bone formation on sand-blasted titanium implants: an experimental study. Biomaterials 17: 1015–1018.

Pilliar RM (1991) Modern metal processing for improved load-bearing surgical implants. Biomaterials 12(2): 95–100.

Pilliar RM (1998) Overview of surface variability of metallic endosseous dental implants: textured and porous surface-structured designs. Implant Dent 7(4): 305–314.

Pohler OE (2000) Unalloyed titanium for implants in bone surgery. Injury 31(Suppl. 4): 7–13.

Pourbaix M (1984) Electrochemical corrosion of metallic biomaterials. Biomaterials 5(3): 122–134. Review.

Redepenning J, Schlessinger T, Burnham S, Lippiello L, Miyano J (1996) Characterisation of electrochemically prepared brushite and hydroxyapatite coatings on orthopaedic alloys. J Biomed Mater Res 30: 287–294.

Rondelli G, Vicentini B, Sivieri E (1997) Stress corrosion cracking of stainless steels in high temperature caustic solutions. Corros Sci 6: 1037–1049.

Sandrini E, Morris C, Chiesa R, Cigada A, Santin M (2005) In vitro assessment of the osteointegrative potential of a novel multiphase anodic spark deposition coating for orthopaedic and dental implants. J Biomed Mater Res 73B(2): 392–399.

Schreckenbach JP, Marx G, Schlottig F, Textor M, Spencer ND (1999) Characterization of anodic spark-converted titanium surfaces for biomedical applications. J Mater Sci Mater Med 10(8): 453–457.

Shabalovskaya SA (1996) On the nature of the biocompatibility and on medical applications of NiTi shape memory and superelastic alloys. Biomed Mater Eng 6(4): 267–289.

Shabalovskaya SA (2002) Surface, corrosion and biocompatibility aspects of nitinol as an implant material. Biomed Mater Eng 12(1): 69–109. Review.

Shieh KJ, Li M, Lee YH, Sheu SD, Liu YT, Wang YC (2006) Antibacterial performance of photocatalyst thin film fabricated by defection effect in visible light. Nanomedicine 2(2): 121–126.

Shirkhanzadeh M (1991) Bioactive calcium phosphate coatings prepared by electrodeposition. J Mater Sci Lett 10: 1415–1417.

Shirkhanzadeh M (1995) Calcium phosphate coatings prepared by electrocrystallization from aqueous electrolytes. J Mater Sci Lett 6: 90–93.

Smith DC (1993) Dental implants: materials and design considerations. Int J Prosthodont 6(2): 106–117.

Southerland DS, Foreshow PD, Allen GC, Brown IT, Williams KR (1993) Surface analysis of titanium implants. Biomaterials 12: 893.

Tisel CL, Goldberg VM, Parr JA (1994) The influence of hydroxyapatite and tricalcium phosphate coating on bone growth into titanium fiber-metal implants. J Bone Joint Surg 76: 139–171.

Vezeau PJ, Keller JC, Wightman JP (2000) Reuse of healing abutments: an in vitro model of plasma cleaning and common sterilization techniques. Implant Dent 9(3): 236–246.

Wagner WC (1992) A brief introduction to advanced surface modification technologies. J Oral Implantol 18(3): 231–235.

Walivaara B, Aronsson BO, Rodahl M, Lausmaa J, Tengvall P (1994) Titanium with different oxides: in vitro studies of protein adsorption and contact activation. Biomaterials 15: 827–834.

Wirtz GP, Brown SD, Kriven WM (1991) Ceramic coating by anodic spark discharge. Mater Manuf Process 6(1): 87–115.

Yerochin AL, Nie X, Leyland A, Matthews A (2000) Characterisation of oxide films produced by plasma electrolytic oxidation of a Ti-6Al4V alloy. Surf Coating Tech 130: 195–206.

Yim CD, Sane SS, Bjarnason H (2000) Superior vena cava stenting. Radiol Clin North Am 38(2): 409–424.

6

CERAMICS

Montserrat Espanol, Román A. Pérez, Edgar B. Montufar,
and Maria-Pau Ginebra

This chapter is devoted to bioceramics, namely those ceramics used in medical applications and therefore intended to be implanted or in contact with tissues and organs. By definition, ceramics are inorganic, nonmetallic materials that are formed by a combination of metallic and nonmetallic elements held together by ionic and covalent bonds. This general definition encompasses a wide variety of compounds, which in turn give rise to a vast variety of properties.

6.1 HISTORICAL PERSPECTIVE

The history of bioceramics provides a good example not only of the progress of the materials themselves but also of the change in mentality and the evolution of the strategies to deal with the problem of materials–body interactions.

Historically, the first attempts to introduce ceramics in clinical applications during the 1960s were directed toward biologically inert ceramics that would be highly resistant to corrosion, so as to minimize their interaction with the body. Alumina and zirconia are good examples of these first-generation ceramics [Hench & Wilson 1993]. Very quickly, however, it was discovered that no material was really inert, and indeed there was a body response to all materials implanted in the body. In fact, the body reacted

Biomimetic, Bioresponsive, and Bioactive Materials: An Introduction to Integrating Materials with Tissues, First Edition. Edited by Matteo Santin and Gary Phillips.
© 2012 John Wiley & Sons, Inc. Published 2012 by John Wiley & Sons, Inc.

to these a priori considered "inert" materials by encapsulating them in a fibrous layer, with this reaction common not only for inert ceramics but also for most metals and plastics. Therefore, this first generation of materials should be better referred to as "nearly inert" or "biostable" materials, in the sense that their properties are maintained after implantation.

In 1969, Hench and coworkers discovered that certain silicate glass compositions could bond to bone [Hench et al. 1972]. This "tissue bonding" response contrasted with the "encapsulation" response, considered in those days to be the universal reaction to any foreign material implanted in the body. Those silicate glasses were termed "bioactive glasses," and their discovery led to the new field of bioactive ceramics. The term "bioactivity" was coined to refer to those materials that had the ability to bond to tissue.

Also in the late 1960s and early 1970s, the quest for better performing implant materials for hard tissue repair resulted in a new concept: bioceramic materials that would mimic natural bone tissue [Levitt et al. 1969; Klawitter & Hulbert 1971]. In fact, bone can be defined as a polymer–ceramic composite, with the ceramic phase nearly 70% in weight. Hydroxyapatite (HA) is the mineral component of bone, and following the biomimetic approach, it was believed that synthetic HA used for bone replacement would be entirely compatible with the body. In fact, later on it was shown that HA was bioactive; that is, it was able to develop a direct bonding with bone, without the formation of a fibrous capsule, the same as the silicate glasses developed by Hench. Since then, the study of HA and other calcium phosphates has attracted much attention and has given rise to several clinical applications for bone regeneration materials.

These ceramics (synthetic HA, bioactive glasses) with direct bone-bonding capabilities belong to the second generation of biomaterials, according to the classification proposed by Hench and described in Chapter 1. The third generation of bioceramics [Hench & Polak 2002] is a further step, which is crucial for regenerative medicine, and which consists of the design of materials where the separate concepts of bioactivity and resorbability converge and progressively disappear while simultaneously guiding the ingrowth of newly formed tissue. Therefore, bioactivity and resorbability will appear as the key features in the biological performance of different ceramic materials; these will be described in more detail in the following sections. In Figure 6.1 are summarized the main current applications of all these types of bioceramics in medicine.

6.2 BIOSTABLE CERAMICS

As introduced earlier in this chapter, the initial application of ceramics as biomaterials was based on the intended limited interaction of corrosion-resistant ceramics with living tissues, even in corrosive environments such as the physiological medium. Although strictly speaking no material can be considered inert, some ceramic materials undergo little chemical change during long-term exposure to the physiological environment and therefore can be called "biostable ceramics." Usually, biostable ceramics with very smooth surfaces form a fibrous layer approximately 1 μm thick [Hench 1998]. However, the thickness of the fibrous layer will depend on several parameters such as the shape, size, surface morphology, porosity, and composition of the implant, as well as the type

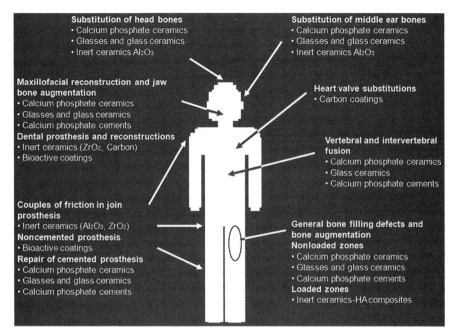

Figure 6.1. Sketch of the main applications of bioceramics in medicine.

of tissue surrounding it and the level of stresses it will have to withstand. Generally, biostable materials are attached to the physiological system through mechanical interlocking.

The most representative biostable bioceramics are aluminum oxide or alumina (Al_2O_3) and zirconium oxide or zirconia (ZrO_2). However, other ceramics such as aluminates, silica nitrates, or pyrolytic carbon (used in autolubricated prostheses and for cardiac valve coatings) have also been implanted and can be considered biostable ceramics. Moreover, it is important to note that some metallic materials such as titanium or stainless steel owe their "inertness" or biostability to the formation of a continuous oxide layer on their surface, which prevents corrosion of the underlying metal by the physiological medium. In fact, this oxide layer is a spontaneous ceramic coating on the metal surface.

6.2.1 Alumina

High purity (>99.5%) alpha aluminum oxide (α-Al_2O_3) is the most used biostable ceramic, and its efficiency has been proven for more than 40 years [Hench 1998]. Alumina has good mechanical properties, especially toughness and strength, which together with its low friction coefficient and excellent wear resistance have made it a good material for articular surfaces in total joint replacements. The excellent wear resistance of alumina has been attributed to its low and valley-type roughness, and to its high surface energy, which gives it a high tendency to adsorb proteins.

Alumina is processed by compacting and sintering powders with granulometries in the range of 2–6 µm. Sintering temperatures are between 1600°C and 1800°C. Fully dense sintered samples can be obtained by mixing the alumina powder with a small amount (less than 0.5 wt %) of magnesium oxide (MgO). The presence of SiO_2 (silica) and alkaline oxides should be avoided because they promote grain growth and impede densification [Hench 1998; Marti 2000]. Mechanical properties such as strength, fatigue resistance, and fracture toughness of the alumina implant depend on the purity and particle size of the raw material and the sintering parameters used in the process (temperature, holding time, heating and cooling rate, atmosphere, and pressure). The International Organization for Standardization (ISO) has provided certain requirements to assure good in vivo performance of these implants. The density should be above 3.90 g/cm^3, and the mean grain size should be kept below 7 µm to ensure good mechanical properties. The bending strength evaluated in saline medium should be higher than 400 MPa [ISO 1994]. Surface finish is very important for the success of alumina implants as wear couples.

6.2.2   Zirconia

Zirconium oxide (ZrO_2) was introduced as an alternative to alumina for femoral ball heads in total hip replacements in the 1980s. The reason was that zirconia had some advantages over alumina ceramics such as higher fracture toughness, due to a transformation toughening mechanism operating in its microstructure [Piconi & Maccauro 1999]. Zirconia can exist as three crystalline phases: monoclinic, which is the stable phase at room temperature; tetragonal, above 1170°C; and cubic, above 2370°C. Usually some metal oxides like yttria (Y_2O_3) or magnesia (MgO) are added to stabilize the tetragonal phase at room temperature. The transformation from tetragonal to monoclinic phase, which could be stress induced, is responsible for the toughening of the ceramic [Marti 2000]. When a crack propagates, the adjacent grains transform from tetragonal to monoclinic, producing a 3–4% increase in volume, which creates a compressive stress field that tends to close the crack, hindering its propagation. Due to this improved toughness, even with its lower wear resistance as compared to that of alumina, zirconia was used as a good alternative for joint replacement and was applied in combination with UHMWPE (ultra high molecular weight polyethylene) as a counterpart to the articular surface.

In fact, zirconia femoral heads were successfully used until 2001, when a series of catastrophic failures started a huge controversial issue on the compound's future as a biomaterial. This resulted in the discontinuation of its clinical use in total hip replacement components [Masonis et al. 2004]. This problem was related to some processing defects, which accelerated the aging or degradation of the material in humid environments caused by the progressive transformation of the metastable tetragonal phase into the monoclinic phase in presence of water [Lawsons 1995]. At present, great efforts are being made to develop alumina zirconia composites as an alternative to monolithic alumina and zirconia [Chevalier 2006].

From another perspective related to the interaction of these materials with tissue, it is worth noting the studies of Nogiwa-Valdez et al. [2006], which successfully introduced calcium and silicon oxides into biostable ceramics in order to add bioactivity to the material. As a result, they observed the in vitro formation of an apatite layer formed after being immersed in simulated body fluid (SBF) [Oyane et al. 2003].

6.3 BIOACTIVE AND RESORBABLE CERAMICS

6.3.1 Basic Concepts

As previously mentioned, the term "bioactive" coined by Hench and coworkers in 1969 refers to a particular response occurring at the tissue–implant interface, that is, when a chemical bond forms across the interface between the implant and the tissue [Hench et al. 1972]. However, the concept had to be expanded to include, under "bioactive ceramics," some calcium phosphate compounds such as HA and a range of glass–ceramic compositions that showed identical bone-bonding characteristics [Hench 1991]. Underlying the phenomena of bioactivity is an ionic exchange between ions from the material's surface and the surrounding environment, leading to formation of a bone-like layer at the material's surface to which bone can bond. Another relevant property that some bioceramics have is resorbability, which refers to the progressive dissolution of the material and its replacement with natural tissues [Hench & Wilson 1993]. The combination of bioactivity and resorbability in a unique implant has set the starting point for the third generation of implants, the new implants designed to actively contribute to tissue regeneration [Hench & Polak 2002].

As will be seen in the following sections, there are many different ways to process ceramic materials, and the choice of one method or another ultimately depends on the required properties for the material at the implant site. While dense and stable (with low or nonexistent resorbability) materials are often sought for sites that require high mechanical strength, highly porous ceramic materials would be needed for tissue regeneration purposes. In general, dense materials are meant to replace lost bone, and porous ones not only to replace but to help the body to regenerate its own lost bone.

Different terms have been defined to describe the interaction of a bioceramic with the host tissue. "Osteoconductivity" refers to the ability of a material to encourage cell adhesion and support their ingrowth. Hence, materials used for this purpose would serve as template or scaffolds to guide bone-forming tissue. The material's role is to provide, at least in the beginning, a stable surface to allow for cell adhesion. Once this is achieved, however, continuous resorption (in vivo dissolution) of the material would be very beneficial, as this will trigger more rapid tissue remodeling and faster bone regeneration. Although HA is a clear prototype of osteoconductive material (bone tissue is osteoconductive and so are its components: HA and collagen [Cornell & Lane 1998]), any of the above-mentioned bioactive ceramics, that is, glass, glass–ceramics, or other calcium phosphates, have shown the same ability [LeGeros et al. 2003]. Osteoconductivity is, in addition, greatly affected by the porosity and degree of pore interconnectivity in the material. Optimum osteoconductivity is achieved when pore size and interconnectivity are adequate to allow cell infiltration (ingrowth) and angiogenesis [Cornell & Lane 1998].

Osteoinductivity, on the other hand, is the ability of a material to induce transformation of progenitor cells to more differentiated bone cells such as osteoblasts. This means that an osteoinductive material has a higher osteogenic potential, even in a nonosseus environment. This property, as mentioned in Chapter 1, has been observed in some bioactive calcium phosphates and glass materials. Rather than being an inherent property of the material, however, it seems to be associated with a critical geometry of the material that permits the concentration of bone morphogenic proteins (BMPs) that are necessary to induce osteoinductivity [Ripamonti et al. 1992]. Although the exact mechanism for osteoinductivity in calcium phosphates is still unknown, Habibovic and coworkers have proposed that during the precipitation process of the bone-like apatite layer on the material's surface, there could be coprecipitation of relevant proteins associated to the differentiation of cells toward the osteogenic lineage [Habibovic et al. 2006]. Instead, osteoinductivity in glasses appears to be intrinsic to its degradation process. It has been found that loading BMPs on these materials would speed up osteoinductivity, leading to accelerated bone regeneration and implant integration [Anselme 2000; Burg et al. 2000]. Therefore, the combination of BMPs on bioactive materials could synergistically enhance integration. At the end of this chapter, we will briefly discuss the new methods adopted for the synthesis of bioceramics that target specific mechanisms of cell adhesion with the goal of enhancing tissue regeneration.

6.3.2 Glasses and Glass–Ceramics

Glasses are noncrystalline materials that can maintain their form and amorphous microstructures if they are held at a temperature that is below their glass transition temperature. The difference as compared to crystalline structures is that in crystals there is a long-range order in which bond angles, interatomic distances, and other variables structure themselves in order to minimize the free energy [Dupree & Holland 1989]. A schematic representation of an amorphous material and a crystalline material is presented in Figure 6.2.

Many studies have been performed since the discovery of bioactive properties in some glass formulations. The most important advantage found in bioactive glasses over other bioactive ceramics is their ability to control a wide range of chemical properties as well as the rate of bonding to tissues. However, this superior ability for bonding to bone contrasts with the poor mechanical properties inherent in the structure of glasses.

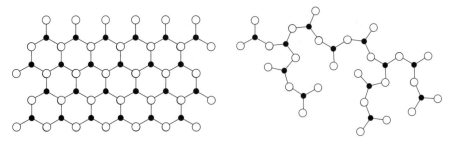

Figure 6.2. Schematic representation of a crystalline (left) and amorphous (right) material.

Glasses suffer from mechanical weakness and low fracture toughness and, as will be seen throughout this chapter, these limitations restrict their applications to coatings, non-load-bearing implants, and bioactive granules. In the next section, the two basic groups of glasses used for medical applications—silicate glasses and phosphate glasses—will be described.

6.3.2.1 *Physicochemical Properties of Bioactive Glasses.*

The most used and well-known bioactive glasses are the silicate-based glasses and, to a lesser extent, the phosphate-based glasses. These are composed of a matrix of SiO_4 or PO_4 tetrahedra, respectively. Silicate dioxide (SiO_2) or phosphorous oxide (P_2O_5) acts as a network-forming oxide in each case. In the case of phosphate glasses, phosphorous oxide is the network-forming agent. Network-modifying oxides can be introduced, which can have either a depolymerization effect, or a crosslinking effect, depending on the number of nonbridging oxygens created within the structure (Figure 6.3). Thus, the nature and properties of the ions introduced within the glass network will modulate the final properties obtained [Navarro et al. 2003a,b], in terms of structure, mechanical properties, and chemical reactivity as summarized in Figure 6.4.

Moreover, depending on the free oxygens available for the branching and the interconnection of the glass, four types of structures appear: Q^0, with no free oxygen; Q^1, or end unit with one free oxygen; Q^2, or middle unit with two free oxygens; and Q^3, or branching units with three free oxygens. The more branching in the glass, the more packed and rigid the structure is. This leads to more soluble glasses as there is no available space in the structure to allocate the network modifiers [Walter et al. 2001].

The network modifiers can be oxides from Na, K, Ca, Mg, Ti, and Fe, with Na and Ca oxides being the most common. The microstructure and composition of the final glass will depend on the composition of the starting raw material and the processing method.

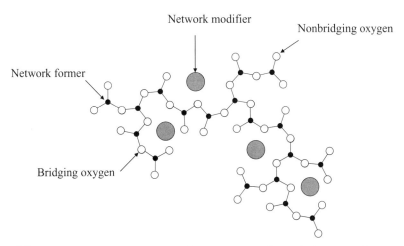

Figure 6.3. Effect of adding modifying oxides in a network formed of network former oxides and bridging oxides.

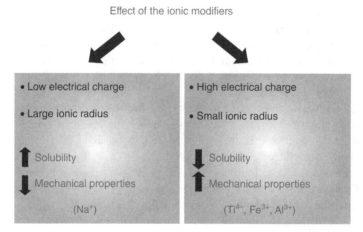

Figure 6.4. Effect of ionic modifiers in the glass properties.

It is important to note that glasses can give rise to glass–ceramics by heat treatment, which results in the formation of various crystalline phases in a glass matrix. This transformation makes glass–ceramics less brittle, since the crystals enclosed in the structure stop crack propagation and in turn increase their mechanical properties [Li et al. 1991; Kasuga & Abe 1999].

6.3.2.2 *Silicate-Based Glasses.*

The first bioactive glass composition developed by Hench was termed Bioglass® 45S5 and was composed of 45% SiO_2, 24.5% Na_2O, 24.4% CaO, and 6% P_2O_5 in weight percent. Studies on this composition have been carried out that changed the composition of the glass while maintaining the 6% P_2O_5 in order to see the effect of different oxides and ratios on the properties. In general, as the calcium-to-phosphorous ratio is decreased, the ability to bond to bone is also decreased. Certain atom substitutions provoke some changes in the properties of glasses, such as the addition of fluorides, which reduces the rate of dissolution. Substituting MgO for CaO or K_2O for Na_2O had little effect. Al_2O_3 and B_2O_3 can modify surface dissolution, but Al_2O_3 is able to inhibit bone bonding. Generally, the addition of multivalent ions makes the glass inactive.

MECHANISM OF BIOACTIVITY. In general, the bioactive nature of glasses and glass–ceramics, that is, their ability to bond to bone, can be assessed in vitro by the formation of a carbonated apatite layer on the surface of the material when immersed in SBF [Kokubo 1998]. The mechanism that leads to the formation of this layer on the Si-based glasses has been carefully studied [Hench & Wilson 1993], and very briefly, it comprises the following stages (Figure 6.5):

1. The fast exchange of network-modifying ions such as Na^+ or K^+ with protons leading to an increase in the interfacial pH.

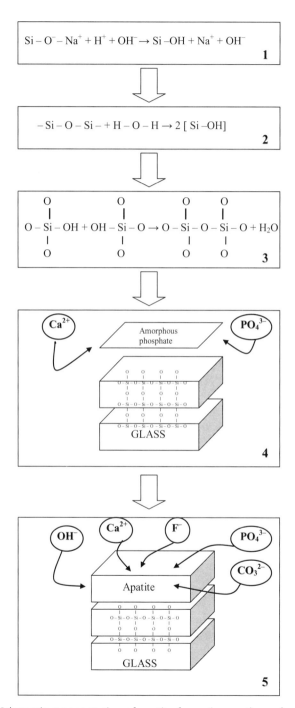

Figure 6.5. Schematic representation of apatite formation on the surface of the glass.

2. The increase in pH promotes network dissolution by breaking the Si–O–Si bonds and is followed by formation of silanol groups (SiOH).

3. The polymerization of the SiO_2-rich layer through condensation of SiOH groups.

4. The migration of Ca^{2+} and PO_4^{3-} groups to the surface of the silica-rich layer, which forms a film rich in CaO–P_2O_5.

5. Crystallization of the amorphous CaO–P_2O_5 film by incorporation of hydroxyls (OH^-) and carbonate (CO_3^{2-}) anions.

6.3.2.3 Phosphate-Based Glasses.

While the structure of phosphate glasses essentially follows that of Si-based glasses, the main feature of phosphate-based glasses over silicate glasses is their high solubility in relation with their structure. In order to suit a particular application, the solubility can be tailored by varying the glass composition. Phosphate glasses have potential applications as biomaterials due to their similar composition to the mineral phase of bone and to their biocompatibility [Navarro et al. 2004a,b]. Moreover, the concern over the long-term stability of silicate glasses means that more interest is being shown in phosphate glasses [Franks et al. 2000]. Phosphate glasses by themselves would not have enough durability to perform in biomedical applications, as they will quickly dissolve. That is why lower solubility compositions are being studied—to see if their stability can be increased by the addition of modifying oxides [Ahmed et al. 2004; Navarro et al. 2005].

Phosphate-based glasses can be classified according to the number of oxides that are present in the glass. Therefore, we can have binary, ternary, and quaternary phosphate glasses.

Modifying the glass composition of a phosphate glass would affect its properties in the same way it does for SiO_2 glasses. For example, the addition of CaO lowers the solubility of the glass while an increase in the amount of K_2O increases its solubility [Knowles et al. 2001]. The presence of the ion Ca^{2+} creates a crosslinked structure between the nonbridging oxygens, which leads to better mechanical properties of the glass [Uo et al. 1998]. Moreover, the addition of TiO_2 strongly increases the stability of the glasses, improving, in turn, their mechanical properties [Navarro et al. 2003a].

6.3.2.4 Processing of Glass and Glass–Ceramics.

Different methods can be used to obtain glasses and glass–ceramics depending on their application. There is an interrelation between processing method, structure, and final properties of the material. Glasses that are obtained from the liquid state are called melt glasses. A very different approach is the sol–gel processing, which allows for the production of glass and glass–ceramics at lower temperatures.

MELT GLASSES. Melt glasses are obtained by heating a mixture consisting of the network former (e.g., SiO_2, silica) with the network modifier oxide (MO) at a temperature well above their melting point (some hundreds of degrees above) for a certain period of time. At such high temperatures, the mixture becomes liquid and, when enough time is allowed, the components become well homogenized. During heating,

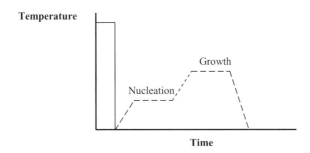

Figure 6.6. Diagram showing the temperature variation for obtaining glass and glass–ceramic. Continuous line corresponds to glass and dotted line glass–ceramic.

the MO molecules break the Si–O–Si bonds of SiO_2. Casting of this liquid will result in a glass or a polycrystalline microstructure depending on the cooling rate and the content of network former in the mixture. To ensure formation of glass, rapid cooling rates are required, as well as enough concentration of network former [Hench & Wilson 1993].

A glass–ceramic can be obtained by post heat treatment of a glass. The heat treatment is carried out in two steps. In the case of Si-based materials, first, the temperature is raised to approximately 450–700°C to produce a high concentration of crystal nuclei. Afterward, the temperature is increased to 600–900°C so that the crystal nuclei can grow (Figure 6.6). The heating rate is a critical parameter and has to be slow enough to avoid the generation of stresses due to volume changes during crystallization. Different crystal sizes can be obtained varying the holding time during the heat-treatment process [de Aza et al. 2007].

SOL–GEL PROCESS. Beginning in the 1990s, sol–gel processing started to be used to produce bioactive glasses to overcome the main limitations of conventional melting processes that required high temperature treatments. The most serious disadvantages of the processing from the melt were the difficulties in controlling the final glass composition due to volatilization of components, in addition to the high cost and limited composition ranges in terms of adequate viscosity of the melt [Li et al. 1991]. Sol–gel processing overcomes the above-mentioned limitations and offers a broader range of new compositions as well as control of bioactivity due to the ease at which compositions and microstructures can be changed throughout the process.

Three methods can be used to obtain sol–gel materials. The first method consists of the gelation of colloidal powders by changing the pH, and the second and third methods depend, respectively, on the controlled hydrolysis and condensation of metal alkoxide precursors (e.g., $Si(OR)_4$, being R $C_xH_{(1+2x)}$ and varying x from 1 to 3) followed by hypercritical drying and drying at ambient temperature [Hench & Wilson 1993]. All three methods involve a long process that aims at generating a three-dimensional (3D) network (or gel) from a suspension of colloidal particles (sol) (Figure 6.7) [Hench & West 1990]. Another advantage of sol–gel processing is the high specific surface area

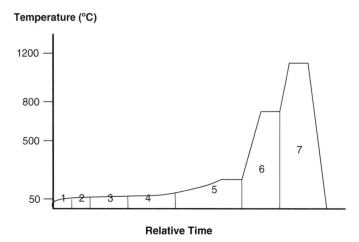

Figure 6.7. Sol–gel process sequence showing temperature dependence. Numbers stand for the different processes: (1) mixing, (2) casting, (3) gelation, (4) aging, (5) drying, (6) dehydration or chemical stabilization, (7) densification.

of the resulting material, which leads to a higher reactivity and increased solubility, thereby improving the material's bioactivity [Vaahtio et al. 2006].

6.3.3 Calcium Phosphates

The term "calcium phosphates" refers to the family of the calcium orthophosphates, pyrophosphates, or dipolyphosphates and polyphosphates. Calcium orthophosphates are salts of phosphoric acid, and thus can form compounds that contain $H_2PO_4^-$, HPO_4^{2-}, and/or PO_4^{3-}. Pyrophosphates $(P_2O_7^{4-})$ and polyphosphates $((PO_3)_n^{n-})$ are distinct from orthophosphates in that their structure contains P–O–P bonds. Among these, calcium orthophosphates containing HPO_4^{2-} and PO_4^{3-} constitute the most biologically relevant compounds. Refer to Elliott 1994 for more detailed information on the structure and chemistry of calcium orthophosphates.

6.3.3.1 *Physicochemistry of Calcium Phosphates.* The inorganic component of bones, teeth, and some pathological calcifications (i.e., dental calculi and urinary stones) has a close resemblance, in terms of composition and structure, to HA, $Ca_{10}(PO_4)_6(OH)_2$ [Armstrong & Singer 1965], from the family of the orthophosphates, which makes this compound a model for mineral tissue. However, strictly speaking, a better model would be nonstoichiometric carbonate-containing HA also known as dahllite or "biological apatite." Besides biological apatites, other orthophosphates such as dicalcium phosphate dihydrate (DCPD), octacalcium phosphate (OCP), and amorphous calcium phosphates (ACPs) have been detected in hard and/or pathological tissues and, together with HA, are being used as potential bioactive and/or resorbable ceramics for bone repair and substitution. Moreover, compounds like tricalcium phosphate (TCP),

TABLE 6.1. Properties of Biologically Relevant Calcium Orthophosphates

Ca/P Ratio	Compound	Abbr.	Chemical Formula	$-\log(K_{sp})$ at 25°C
0.5	Monocalcium phosphate monohydrate	MCPM	$Ca(H_2PO_4)_2 \cdot H_2O$	1.14
0.5	Monocalcium phosphate anhydrous	MCPA	$Ca(H_2PO_4)_2$	1.14
1.00	Dicalcium phosphate dihydrate, "Brushite"	DCPD	$CaHPO_4 \cdot 2H_2O$	6.59
1.00	Dicalcium phosphate anhydrous, "Monetite"	DCPA	$CaHPO_4$	6.90
1.33	Octacalcium phosphate	OCP	$Ca_8(HPO_4)_2(PO_4)_4 \cdot 5H_2O$	96,6
1.50	α-Tricalcium phosphate	α-TCP	$Ca_3(PO_4)_2$	25.5
1.50	β-Tricalcium phosphate	β-TCP	$Ca_3(PO_4)_2$	28.9
1.5–1.67	Calcium-deficient hydroxyapatite	CDHA	$Ca_{10-x}(HPO_4)_x(PO_4)_{6-x} \cdot (OH)_{2-x}$ $(1 < \times < 0)$	~85.1
1.67	Hydroxyapatite	HA	$Ca_{10}(PO_4)_6(OH)_2$	116.8
2.00	Tetracalcium phosphate, "Hilgenstockite"	TTCP	$Ca_4O(PO_4)_2$	38–44
1.2–2.2	Amorphous calcium phosphate	ACP	$Ca_x(PO_4)_y \cdot nH_2O$	–

and tetracalcium phosphate (TTCP), which are not found in mineralized tissues (except for the magnesium-containing form of beta-TCP), have found application as resorbable compounds because of their ability to hydrolyze into other most stable calcium orthophosphates [Elliott 1994; Dorozhkin & Epple 2002]. Table 6.1 presents the most relevant calcium orthophosphates used for biological and medical applications.

As mentioned previously, bone mineral has an apatitic structure. The general chemical formula to describe apatites is $[A(1)]_4[A(2)]_6(BO_4)_6X_2$, where A represents generally bivalent cations distributed on two crystallographic sites (1 and 2 in the chemical formula), and BO_4 and X are, respectively, trivalent and monovalent anions. The apatite family crystallizes in the hexagonal system ($P6_3/m$) [Elliott 1994]. A unique characteristic of the apatite structure is its hospitability in allowing substitutions of many different ions in the lattice. When these substitutions lead to the introduction of ions with different charges, as it does for carbonate, CO_3^{2-}, vacancies are also introduced to retain electroneutrality. This property imparts in apatites the ability to adapt to the biological function of different tissues [Cazalbou et al. 2004]. Incorporation of different ions and vacancies in the crystal structure of HA renders apatites with different chemical compositions and different properties in terms of crystallinity, thermal stability, and solubility.

Since reactivity and resorbability relate to the rate of dissolution of a particular compound, knowing the solubility product for the different orthophosphates will be a good predictor of their behavior in aqueous or body fluid environments. Table 6.1 includes the logarithm of the solubility product [Dorozhkin & Epple 2002]. While the forming method and exact stoichiometry will also have an effect on solubility, the generally accepted order of solubility in physiological conditions is

$$ACP > DCPD > TTCP > \alpha\text{-}TCP > \beta\text{-}TCP \gg HA.$$

The limited solubility of HA in contact with aqueous solutions, compared to the other calcium orthophosphates, is not surprising, as it is the most stable calcium phosphate compound at pH levels above 4.2. Although after implantation the pH may locally drop to a pH around 5 in the region of tissue damage, it would eventually return to pH 7.4 over a short period of time. Even under these conditions, HA is still the stable phase, but below a pH of 4.2, DCPD is the most stable compound. In vivo, the superior solubility of ACP, DCPA, TTCP, and TCP results in their dissolution/resorption leading to an increase in the levels of calcium and phosphate ions, which in turn triggers precipitation of a most stable compound (e.g., apatite at a pH above 4.2). Complications in the development of resorbable orthophosphates come from the maintenance of strength and stability at the interface during the degradation process and also matching the resorption rates with that of bone tissue ingrowth.

Moreover, apart from the solubility degree dictated by the solubility product for each compound, features such as high surface area availability (powder > porous solid > dense solid), smaller crystal size, low crystallinity, high level of crystal imperfection, and the presence of specific ionic substitutions would add to the compound's solubility [Ratner et al. 1996]. Therefore, resorbability is achieved in bioactive HA by either reducing its crystal size or augmenting crystal imperfection or by mixing HA with another more resorbable compound. All these features come into play during the synthesis and further processing of the materials and will be detailed in the following sections.

MECHANISM OF BIOACTIVITY. The great advantage of bioactive materials like HA is their ability to form an apatite bone-like layer on their surface to which bone bonds under in vivo and in vitro (i.e., SBF) conditions [Oyane et al. 2003]. The exact mechanism to explain the growth of this apatite layer is not yet fully understood. On the one hand, it seems that, following the classical nucleation theory, either an increase in supersaturation of the solution or lowering the interfacial energy of HA would increase the rate of nucleation that is required for spontaneous crystal growth (at a given temperature and supersaturation). However, the low solubility of HA in water does not help to increase supersaturation much, nor does its low interfacial energy. Studies by Nancollas and coworkers [Nancollas & Wu 2000] suggest that OCP and DCPD, both having much lower interfacial tension than HA, could be first nucleated on HA and act as precursors for HA mineralization. However, another potential precursor is ACP. Kim et al. [2004] have explained mineralization of bone-like apatite on HA through formation of ACP as a consequence of a sequence of electrostatic interactions with the HA surface.

6.3.3.2 *Processing of Calcium Orthophosphates.* The objective for
calcium orthophosphates processing—and essentially any bioceramic—is to shape a
material so that it can perform a specific function. This requires making solid objects,
coatings, or granules. Granules are essentially used for space filling and regeneration
of tissues. Coatings on metallic substrates provide mechanical support while allowing
tissue bonding and corrosion protection. Bulk ceramics are mainly used for replacement
and augmentation of tissue in low load-bearing applications. As in glass and glass–
ceramics, the inherent brittleness of ceramics reduces the field applications of calcium
orthophosphates as a bulk ingredient.

Calcium phosphate processing can be carried out either following the traditional
high temperature ceramic conformation methods, or alternatively by following processing methods more similar to the biological routes, that is, low temperature and solution-
mediated routes. The two strategies are described below.

(a) High temperature processing of calcium orthophosphates

It is usually performed to obtain:

(1) dense and porous bulk samples, and
(2) coatings on metallic substrates.

There are two main strategies for making bulk ceramic samples with a specific shape:
through preforming the shape from powders by compaction in a mold or by casting a
powder suspension from the liquid state into a mold (slip casting). Both ways require
an additional firing or sintering (heat treatment) step to consolidate the bulk in order
to reduce porosity and improve strength. "Dense" or "porous" bulk samples can be
obtained, depending on the presence of micro- or macro- (above 100 μm) porosity
[LeGeros 1988]; while the former is inherent to the processing method and conditions
(a maximum of 5% of microporosity by volume is accepted in dense samples [Hench
& Wilson 1993]), the second is intentionally introduced with the help of pore-creating
agents such as naphthalene, paraffin, peroxide, and so on. As mentioned previously, the
rationale for synthesizing porous bulk samples stems from their ability to allow angiogenesis and growth of natural tissue inside the pores while helping to mechanically
stabilize the implant. It is important to emphasize that the tissue response to chemically
identical samples with different porosities (i.e., one being dense while the other porous)
is different because of the opportunity for ingrowth in the porous material. Increasing
the degree of porosity and interconnectivity would promote tissue ingrowth (and therefore its integration) but would make the implants mechanically weaker. In the case of
the dense materials, osteoconduction would be limited on its surface.

The starting raw material for the processing of either dense or porous samples is
in powder form, and it is usually HA, or beta-TCP whenever higher resorbability is
required. A mixture of both phases can also be used, usually termed "biphasic calcium
phosphates" (BCPs). Adjusting the relative proportions of each phase allows tuning the
resorption rate of the ceramic [LeGeros 1988].

HA powders can be sintered (heated below their melting point) at temperatures ranging from 1000°C to 1200°C following compaction of the powder into a desired shape [Jarcho 1981]. High densities at lower processing temperatures can be achieved by applying pressure while heating the powder, that is, hot pressing (HP), hot isostatic pressing, (HIP), spark plasma sintering (SPS), and so on.

An obvious limitation for the powder pressing techniques (HP, SPS, etc.) is that only limited and simple shapes can be made. When different shapes are required, alternative processing techniques need to be used. The most common ones applied to produce HA bulk samples are slip casting [Sadeghian et al. 2005] and tape casting [Krajewski et al. 1982], among others [Suchanek & Yoshimura 1998]. In slip casting, a suspension containing the HA powder (slurry) is poured into a mold with the desired shape, water from the slip is adsorbed into the mold leaving behind the solid piece, which is subsequently fired. Tape casting is used to produce HA sheets by pouring the slip onto a flat surface, and a doctor blade is used to spread uniformly the slip into the thin tape. Consolidation is accomplished by firing.

A completely different application from the non-load-bearing dense/porous bulk calcium phosphates arises when using calcium phosphate (ideally HA) coatings. The concept of coating HA on a metallic implant combines the mechanical advantages of the metal with the excellent biocompatibility and bioactivity of HA, making those implants suitable for high load-bearing applications. The HA coatings fulfill several functions: to provide stable fixation of the implant to the bone, decrease the release of metal ions from the implant to the body, and shield the metal surface from environmental attack. Several techniques have been used to coat the implant with HA metallic substrates [Suchanek & Yoshimura 1998; de Groot et al. 1998], but by far, the most popular technique in terms of cost-effectiveness, efficiency, and reproducibility is plasma spraying. The plasma spray process basically involves the spraying of molten or semimolten particles onto a substrate. The material, in the form of powder, is injected into a very high temperature plasma jet (10,000–15,000K) and is accelerated (1000–1500 m/s) toward the metallic substrate where it rapidly cools upon impingement, forming a coating (with a quenching rate $> 10^6$ K/s). A common problem encountered in plasma spraying is the decomposition of the HA powder to other resorbable phases such as ACP, which often forms at the coating–substrate interface. Their fast dissolution in aqueous environments can compromise the coating's stability in vivo and be the cause of implant failure [Sun et al. 2001].

(b) Low temperature processing of calcium orthophosphates

It is usually performed to obtain:

(1) calcium phosphate cements (CPCs), and
(2) biomimetic coatings.

Generally speaking, calcium orthophosphate cements are mixtures of one or more of the following powders: TCP, DCPD, DCPA, MCPM, OCP, or TTCP with water or a calcium- or phosphate-containing solution. For certain combinations, the mixture trans-

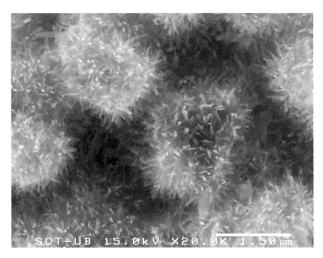

Figure 6.8. Detail of the fine needle-like structure on the surface of a cement.

forms into calcium-deficient hydroxyapatite (CDHA) during setting, forming a porous body even at 37°C [Driessens et al. 1998; Fernández et al. 1999a,b]. The setting time can be as short as a few minutes. The consolidation of the material is attained not by sintering as in traditional ceramics but through a low-temperature setting reaction, and specifically through the entanglement of the crystals produced as a result of a dissolution–precipitation reaction [Ginebra et al. 1997]. This approach renders several advantages to the material: first, the cementing reaction takes place at low temperatures and therefore the final product is a low-temperature precipitated hydroxyapatite (PHA); second, chemically, the cement is similar to biological apatites; and third, the cements possess a much higher specific surface than sintered HA samples. The detail of the fine crystal surface of a cement is shown in Figure 6.8 All these factors contribute to giving the material a higher reactivity, as compared to a ceramic HA. Moreover, as for the sintered HA ceramic, porogen agents or foaming agents can be added during the making of the cement to add macroporosity to the material [Almirall et al. 2004; Ginebra et al. 2007; Montufar et al. 2010; Perut et al. 2011]. In addition, it will be highlighted in the next section how the low processing temperature in cements allows the introduction of proteins [Kamegai et al. 1994; Otsuka et al. 1994a; Blom et al. 2000; Knepper-Nicolai et al. 2002] or different kinds of drugs into the material.

The physicochemical properties of the cement such as setting time, porosity, and mechanical behavior strongly depend on the cement formulation, the presence of additives, the initial particle size of the raw material, and the reaction conditions (temperature and pH) [Ginebra et al. 1995, 2004]. Some of the great advantages of cements are that they can be easily shaped and handled in paste form and that they can be made injectable and used in minimally invasive surgical techniques. These features make cements more attractive for filling bone defects as opposed to bulk samples, which are more difficult to shape, and more suitable than granules, which are difficult to keep in place. However, at present, the main drawback for bone cements is their relatively poor mechanical strength, which limits their applicability to non- or low-load-bearing applications.

From the biological/clinical point of view, CPCs have proven to have unique properties. In fact, it is well known that they are biocompatible, bioconductive, and that they support constant remodeling in vivo [Ooms et al. 2003]. Indeed, since CPCs were discovered by Brown and Chow in 1983 [Brown & Chow 1986; Chow 1991], their use as materials for bone regeneration has been increasing in different fields such as ortho-pedic surgery, dentistry, maxillofacial surgery, and reconstructive surgery [Wise et al. 2000; Bohner 2001].

Biomimetic coatings is a rather new approach proposed by Tadashi Kokubo in the late 1980s to coat metallic substrates or other surfaces with bone-like apatite [Abe et al. 1990; Gil et al. 2002], and this approach offers the most promising alternative to plasma spraying and other coating methods (either high- or low-temperature processes) [Epinette & Manley 2004]. The coating is continuously formed upon soaking the sub-strate, previously chemically treated to induce apatite formation, in SBF. SBF with ion concentrations nearly identical to those of human blood plasma is supersaturated with respect to HA, which promotes apatite nucleation on the treated substrates. While this technique results in formation of a uniform dense layer at low temperature (e.g., room temperature) and enables the coating of complex substrates, it presents some drawbacks with respect to conventional plasma spraying. These drawbacks include the long pro-cessing times required to achieve thick layers and the fact that the process needs replenishment of solutions and constant conditions [Kamitakahara et al. 2007]. The classical biomimetic coating normally requires an immersion period of about 7–28 days, with replenishment of SBF solution. In recent years, efforts have been made to speed up this process to increase its practical utility [Tas & Bhaduri 2004; Aparicio et al. 2007]. Similarly, as was described for cements, the bone-like apatite crystals that result from the biomimetic coating have high bioactivity and good resorption charac-teristics. Moreover, the low temperature of the process allows, as with cements, incor-poration of bone growth stimulating factors to induce bone formation.

6.3.4   New Trends in Bioactive and Resorbable Materials Integration

While evidence of the wide acceptance of bioceramics in our lives is the large number of commercial brands dedicated to their fabrication and commercialization, the market is continuously moving toward development of newer products that will further enhance the interaction of bioceramics with the body. Table 6.2 summarizes some commercial-ized bioceramics for bone-filling applications, classified by the type and component/origin.

The aim of this section is to introduce to the reader the new approaches in the bioceramics field that target specific implant–body responses to further accelerate the material's integration and tissue regeneration.

Throughout this chapter, different bioceramics have been introduced that are designed to respond "specifically" in the body: the biostable resulted in the least inter-action with the body, the bioactive reacted in the body by allowing direct chemical

TABLE 6.2. Bone-Filling Bioceramic Brands Classified by Type and Component/Origin

Type	Component/Origen	Product Name	Company	Country
Calcium phosphates	TCP	Biobase	BioVision	Germany
		Betabase		
		chronOS	Mathys	Switzerland
	HA	ProOsteon	Interpore	USA
		Pyrost	Stryker Howmedia	USA
		Calcitec	Sulzer	USA
		Synthocer		Germany
		Osteogen	Impladent	USA
	BCP	MBCP	Biomatlante	France
		Osteosynt	Eico Ltds	Brazil
		Triosite	Zimmer	France
		Ceratine	NGK Spark Plug	Japan
Bioactive glasses and glass–ceramics	Bioglass 45S5	PerioGlas NovaBone Consil	Novabone	USA
	Glass–ceramics	Ceravital	Leitz GmbH	Germany
		Cerabone A/W	Nippon electric	Japan
		Ilmaplant-L1	–	–
		Bioverit (I and II)	VITRON GmbH	Germany
Hybrid materials	PE–HA	HAPEX	Richards Medical	USA
		Fin-Biotape	Fin Ceramics	Italy
Naturals	Coral	Interpore Prosteon	Interpore	USA
	Bovine bone	BioOss	Geitslich	Switzerland
		Osteograf-N	CeraMed	USA
		Endobon	Merck GmbH	Germany
		Nukbone	Biocriss	México
Calcium phosphate cements		α-BSM	ETEX	UK
		Bone Source	Othofix	USA
		Biopax	Mitsubishi	Japan
		Biobone	Merck GmbH	Germany
		Mimix	BioMet	USA
		Norian (SRS and CRS)	Norian	USA
		Cementek	Teknimed	France
		Biocement D	Merck GmbH	Germany
		Fracture Grout	Norian	USA

bonding to bone, the resorbable was meant to progressively dissolve, and the combination of both bioactive and resorbable has become the new generation of materials designed to help the body to heal itself. Moreover, seeking regeneration of functional tissue requires stimulation of specific cell responses such as proliferation, differentiation, and extracellular matrix production and organization, which implies that there has to be some kind of recognition between the material and the environment. Bone tissue engineering and drug delivery are two good examples of material-tissue recognition and will be further developed.

(a) Ceramics in tissue engineering

Bone tissue engineering is a new field of research that applies the principles of engineering and the life sciences for the development of materials able to induce the growth of new functional tissue. The ideal in the development of bone tissue engineering is to obtain bones (sections or complete ones) that are personalized and scaled up to the size of the damaged bone that has to be healed, which could allow, for example, for the recovery of amputated limbs of the body, such as from accidents.

The basic approach is to fabricate a matrix (porous material or 3D scaffold) to serve either as mechanical support for cell attachment or as template or carrier of other agents [Burg et al. 2000; Kneser et al. 2002]. Different materials of varying rigidity can be used as scaffolds depending on their application: rigid, flexible, or gel-like matrices. Bioactive glasses and ceramics are good candidates for the fabrication of bone tissue engineering scaffolds due to their osteoconductive and/or osteoinductive properties. Moreover, bioceramic scaffolds provide engineered tissue with high mechanical stability and definite shape. They are basically made of HA-based calcium–phosphate compounds [Habraken et al. 2007], BCPs [Yuan et al. 2006], and bioactive glasses [Zhong & Greenspan 2000; Navarro et al. 2004a,b] owing to their high biocompatibility and osteoconductive properties. A high degree of macroporosity and porous interconnection is required to allow cells to infiltrate. The presence of microporosity is highly beneficial as it improves fluid penetration inside the scaffold and provides the nutrients necessary for the cells. An aspect that limits the application of ceramics as scaffolds is their low degradability in vivo [Habraken et al. 2007]. Although ceramics are sought for their rigidity, a controlled resorbability closely matching the rate of tissue ingrowth would enhance bone remodeling, which is so necessary for tissue integration. In the previous section, two possible ways of introducing macroporosity in sintered ceramics and in cements through the addition of porogen agents were explained. In terms of higher specific surface area, cements would be most attractive as opposed to sintered porous bodies. The high specific surface area due to macro- and nano/microporosities and the particular microstructure of entangled crystals in cements (refer to Figure 6.8) account for their improved degree of porosity and enhanced resorbability as compared to the sintered samples. In the same way, glasses prepared by the sol–gel method are more suitable as they incorporate more porosity than conventional glass processing [Zhong & Greenspan 2000]. The so-called microcarriers are a different type of scaffold, first developed by Van Wezel in 1967 to seed cells on the scaffolds [Van Wezel 1967]. They are small beads between 100 and 300 μm in diameter, and as they allow cells to

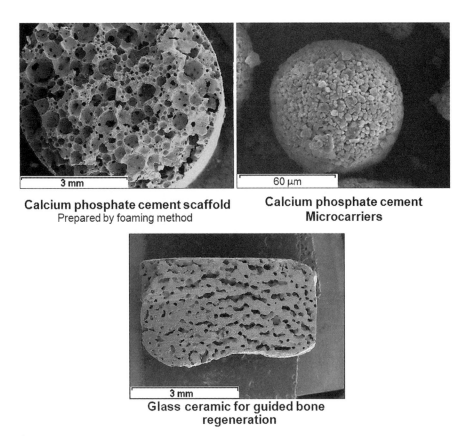

Calcium phosphate cement scaffold
Prepared by foaming method

Calcium phosphate cement Microcarriers

Glass ceramic for guided bone regeneration

<u>Figure 6.9.</u> Bioactive and resorbable ceramics with potential applications as bone tissue engineering scaffolds. Left top: macroporous calcium phosphate cement scaffold. Right top: calcium phosphate cement microcarrier. And bottom: bioactive glass–ceramic for bone grown.

attach on them, they facilitate the homogeneous distribution of cells throughout the replaced tissue. This is sometimes difficult in certain porous scaffolds and often lead to limited cell penetration. In Figure 6.9 are shown different bioactive and resorbable ceramics with potential applications as bone tissue engineering scaffolds.

However, to further add to the degradability of the ceramic, while also reducing the overall brittleness and rigidity of the scaffold, polymers are added to create composites. Two different constructions are possible: adding the ceramic to the polymer (polymer-based composite) or adding the polymer to the ceramic (ceramic-based composite). These polymers can be as varied as synthetic materials designed with specific properties (i.e., degradation rate); to protein-based polymers such as collagen and gelatin among others, with the aim of further mimicking natural tissue; to other natural sources of polymers like chitin, chitosan, alginate hylauronan, and cellulose with a wide range of degradability. The most employed synthetic polymers are polylactic acid (PLA, PLLA), polyglycolic acid (PGA), poly-(D/L-lactic-co-glycolic acid (PLGA), and

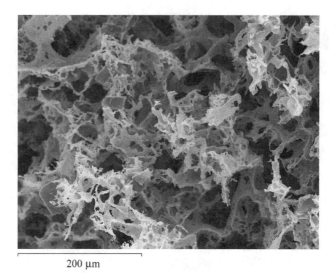

200 μm

Figure 6.10. Microstructure of a composite PLA (polylactic acid) phosphate glass scaffold.

poly-e-caprolactone (PCL). The degradation rate of these polymers decreases in the following order: PGA > PLGA > PLA > PCL [Habraken et al. 2007]. An example of the microstructure of a composite PLA phosphate glass scaffold is shown in Figure 6.10 [Navarro et al. 2004a]. Another composite of great interest used for engineering different types of surfaces is HA- and collagen-based composites. The increasing interest in these types of composites arises not only from their composition and structural similarities to natural bone but also from their unique functional properties in terms of a larger surface area and superior mechanical strength (when compared to their individual constituents) [TenHuisen et al. 1995; Murugan & Ramakrishna 2005]. Processing of the HA–collagen composites is still under study to further enhance the interfacial bonding between both components [Okazaki et al. 1990] while maintaining the nanocrystal size of HA that is necessary for fast bone remodeling [Marouf et al. 1990; Green et al. 2002].

Next to the scaffold in bone tissue engineering is the possibility of incorporating cells and growth factors to induce and accelerate bone ingrowth. There are a large number of growth factors to specifically stimulate proliferation and/or differentiation of osteogenic cells (osteogenesis is the ability to form new bone tissue), but by far, BMPs are the most popular osteoinductive factors [Hubbell 1999]. Usually, incorporation of BMPs is simply done by immersion of the scaffold in a BMP solution, with proteins adsorbing onto the material. A chief disadvantage of this approach is the high amount of proteins that have to be loaded to guarantee the appropriate tissue response. The problem appears from the high binding of the protein to the material, which impedes their release [Habraken et al. 2007]. In the case of seeding cells on the scaffolds, different types of osteogenic cells can be used, but it is currently still unknown which type of cells are best for engineering bone tissue [Kneser et al. 2002].

(b) Ceramics in drug delivery systems

A very new field that is developing for ceramics is their application as delivery systems to carry drugs and/or genes. Drug carriers are used as vehicles to improve the delivery and the effectiveness of drugs. At present, most carriers are made of biodegradable polymers. However, in specific applications in the musculoskeletal system, bioactive ceramics such as calcium phosphates, glass, and glass–ceramics can act as multifunction materials by filling or substituting bone tissue, promoting surface reactivity, and simultaneously acting as punctual delivery systems [Ginebra et al. 2006a,b]. The main features required for drug carriers are large surface areas to allow for enough loading of the drug, and a controlled delivery of the drug able to maintain a concentration that is nontoxic (too fast a release) yet effective (if the release is too slow, the drug may never reach adequate efficacy levels) [Arkfield & Rubenstein 2005]. The same processing routes described earlier in this chapter and employed for the development of porous tissue engineered scaffolds can be used in the development of carriers. But of special interest is the synthesis of carriers using CPCs or the sol–gel method in glasses, as they maximize surface area and offer the possibility of incorporating the drug homogeneously throughout the whole volume, which prolongs its release time [Ellerby et al. 1992; Ginebra et al. 2006a,b]. The addition of drugs on an already processed solid calcium phosphate or glass carrier would not be as effective as in cements and sol–gel glasses, not only because the drug would only mostly adsorb on the material's surface, but also because the release pattern of the drug would be too short. While the release pattern in porous sintered ceramics consists of an initial burst followed by a specific release dependent on the type of drug and type of surface, the release characteristics of cements are controlled by the diffusion of the drug through the pores and are influenced by the material characteristics [Otsuka et al. 1994a,c]. Microcarriers can also be used for drug delivery applications, and they offer the possibility of being injected by mixing them with a gel [Flautre et al. 1996; Komlev et al. 2002; Freiberg & Zhu 2004]. Among the different drugs that have been incorporated in bioceramics are antibiotics [Yu et al. 1992; Bohner et al. 1997], anticancer drugs [Otsuka et al. 1994c], anti-inflammatories [Otsuka et al. 1994b; Ginebra et al. 2001], bone growth factors [Kamegai et al. 1994; Blom et al. 2000], proteins [Otsuka et al. 1994a; Knepper-Nicolai et al. 2002], and hormones and genes [Ginebra et al. 2006b].

EXERCISES/QUESTIONS FOR CHAPTER 6

1. Provide examples of biostable ceramics and their biocompatibility in specific clinical applications.
2. Describe the main physicochemical features of bioactive ceramics.
3. Which are the main types of bioactive glasses and which is their most typical clinical application?
4. Describe a typical sol–gel process for bioglass synthesis.
5. Provide at least two examples of calcium phosphate and critically assess their rate of resorption in vivo in the light of the bone repair timescale.

6. Highlight the main types of engineering applied to ceramics during the manufacturing of biomedical implants.

7. Critically discuss the biomimetic and bioactive potential of hydroxyapatite and tricalcium phosphates.

8. Critically discuss the main requirements for ceramic scaffolds and their use in tissue engineering.

9. Provide examples of composite materials based on biodegradable polymers and ceramics.

10. Highlight the main physicochemical properties of ceramics that stimulate osteoblast adhesion, proliferation, and differentiation.

REFERENCES

Abe Y, Kokubo T, Yamamuro T (1990) Apatite coatings on ceramics, metals and polymers utilizing a biological process. Journal of Materials Science: Materials in Medicine 1: 233–238.

Ahmed I, Lewis M, Olsen I, Knowles JC (2004) Phosphate glasses for tissue engineering: Part 1. Processing and characterisation of a ternary-based P2O5-CaO-Na2O glass system. Biomaterials 25: 491–499.

Almirall A, Larreq G, Delgado JA, Martínez S, Planell JA, Ginebra MP (2004) Fabrication of low temperature macroporous hydroxyapatite scaffolds by foaming and hydrolysis of an alpha-TCP paste. Biomaterials 25: 3671–3680.

Anselme K (2000) Osteoblast adhesion on biomaterials. Biomaterials 21: 667–681.

Aparicio C, Manero JM, Conde F, Pegueroles M, Planell JA, Vallet-Regi M, Gil FJ (2007) Acceleration of apatite nucleation on microrough bioactive titanium for bone-replacing implants. Journal of Biomedical Materials Research 82A: 521–529.

Arkfield DG, Rubenstein E (2005) Quest for the Holy Grail to cure arthritis and osteoporosis: emphasis on bone drug delivery systems. Advanced Drug Delivery Reviews 57: 935–944.

Armstrong WD, Singer L (1965) Composition and constitution of the mineral phase of bone. Clinical Orthopaedics and Related Research 38: 179–190.

Blom EJ, Klein-Nulend J, Klein CPAT, Kurashina K, van Waas MA, Burger EH (2000) Transforming growth factor-beta 1 incorporated during setting in calcium phosphate cement stimulates bone cell differentiation in vitro. Journal of Biomedical Materials Research 50: 67–74.

Bohner M (2001) Physical and chemical aspects of calcium phosphates used in spinal surgery. European Spine Journal 10: S114–S121.

Bohner M, Lemaitre J, Van Landuyt P, Zambelli P, Merkle H, Gander B (1997) Gentamicin-loaded hydraulic calcium phosphate bone cements as antibiotic delivery system. Journal of Pharmaceutical Sciences 86: 565–572.

Brown WE, Chow LC (1986) U.S. patent 4,612,053.

Burg KJL, Porter S, Kellam JF (2000) Biomaterials developments for bone tissue engineering. Biomaterials 21: 2347–2359.

Cazalbou S, Combes C, Eichert D, Rey C (2004) Adaptative physico-chemistry of bio-related calcium phosphates. Journal of Materials Chemistry 14: 2148–2153.

Chevalier J (2006) What future for zirconia as a biomaterial. Biomaterials 27: 535–543.

Chow LC (1991) Development of self-setting calcium phosphate cements. Journal of the Ceramic Society of Japan (International Edition) 99: 927–935.

Cornell CN, Lane JM (1998) Current understanding of osteoconduction in bone regeneration. Clinical Orthopaedics and Related Research 355: S267–S273.

de Aza PN, de Aza AH, Pena P, de Aza S (2007) Bioactive glasses and glass-ceramics. Boletín de la Sociedad Española de Cerámica y Vidrio 46: 45–55.

de Groot K, Wolke JGC, Jansen JA (1998) Calcium phosphate coatings for medical implants. Proceedings of the Institution of Mechanical Engineers. Part H, Journal of Engineering in Medicine 212: 137–147.

Dorozhkin SV, Epple M (2002) Biological significance of calcium phosphates. Angewandte Chemie (International Edition) 41: 3130–3146.

Driessens FCM, Planell JA, Boltong MG, Khairoun I, Ginebra MP (1998) Osteotransductive bone cements. Journal of Engineering in Medicine 212(6 Part H): 427–435.

Dupree R, Holland D (1989) MAS NMR: A New Spectroscopic Technique for Structure Determination in Glasses and Ceramics. New York: Chapman and Hall.

Ellerby LM, Nishida CR, Nishida F, Yamanaka SA, Dunn B, Valentine JS, Zink JI (1992) Encapsulation of protein in transparent porous silicate glasses prepared by the sol-gel method. Science 255: 1113–1115.

Elliott JC (1994) Structure and Chemistry of the Apatites and Other Calcium Orthophosphates. Amsterdam, The Netherlands: Elsevier.

Epinette JA, Manley MT (2004) Fifteen Years of Clinical Experience with Hydroxyapatite Coatings in Joint Arthroplasty. France: Springer-Verlag.

Fernández E, Gil FJ, Ginebra MP, Driessens FCM, Planell JA (1999a) Calcium phosphate bone cements for clinical applications: Part II: Precipitate formation during setting reactions. Journal of Materials Science: Materials in Medicine 10: 177–183.

Fernández E, Gil FJ, Ginebra MP, Driessens FCM, Planell JA, Best SM (1999b) Production and characterization of new calcium phosphate bone cements in the $CaHPOv_4$–α-$Ca_3(PO_4)_2$ system: pH, workability and setting times. Journal of Materials Science: Materials in Medicine 12: 223–230.

Flautre B, Pasquier G, Blary MC, Anselme K, Hardouin P (1996) Evaluation of hydroxyapatite powder coated with collagen as an injectable bone substitute: microscopic study in rabbit. Journal of Materials Science: Materials in Medicine 7: 63–67.

Franks K, Abrahams I, Knowles JC (2000) Development of soluble glasses for biomedical use: Part I. In vitro solubility measurement. Journal of Materials Science: Materials in Medicine 11: 609–614.

Freiberg S, Zhu XX (2004) Polymer microspheres for controlled drug release. International Journal of Pharmaceutics 282: 1–18.

Gil FJ, Padrós A, Manero JM, Aparicio C, Nilsson M, Planell JA (2002) Growth of bioactive surfaces on titanium and its alloys for orthopaedic and dental implants. Materials Science and Engineering C 22(1): 53–60.

Ginebra MP, Boltong MG, Fernández E, Planell JA, Driessens FCM (1995) Effect of various additives and temperature on some properties of an apatitic calcium phosphate cement. Journal of Materials Science: Materials in Medicine 6: 612–616.

Ginebra MP, Fernandez E, De Maeyer EA, Verbeeck RM, Boltong MG, Ginebra J, Driessens FC, Planell JA (1997) Setting reaction and hardening of an apatitic calcium phosphate cement. Journal of Dental Research 76: 905–912.

Ginebra MP, Rilliard A, Fernández E, Elvira C, San Román J, Planell JA (2001) Mechanical and rheological improvement of a calcium phosphate cement by the addition of a polymeric drug. Journal of Biomedical Materials Research 57: 113–118.

Ginebra MP, Driessens FCM, Planell JA (2004) Effect of the particle size on the micro and nanostructural features of a calcium phosphate cement: a kinetic analysis. Biomaterials 25: 3453–3462.

Ginebra MP, Traykova T, Planell JA (2006a) Calcium phosphate cements as bone drug delivery systems: a review. Journal of Controlled Release 113(2): 102–110.

Ginebra MP, Traykova T, Planell JA (2006b) Calcium phosphate cements: competitive drug carriers for the musculoskeletal system? Biomaterial 27(10): 2171–2177.

Ginebra MP, Delgado JA, Harr I, Almirall A, Del Valle S, Planell JA (2007) Factors affecting the structure and properties of an injectable self-setting calcium phosphate foam. Journal of Biomedical Materials Research. Part A 80A(2): 351–361.

Green D, Walsh D, Mann S, Oreffo ROC (2002) The potentials of biomimesis in bone tissue engineering: lessons from the design and synthesis of invertebrate skeletons. Bone 30: 810–815.

Habibovic P, Sees TM, van den Doel MA, van Blitterswijk CA, de Groot K (2006) Osteoinduction by biomaterials-physicochemical and structural influences. Journal of Biomedical Materials Research 77A: 747–762.

Habraken WJEM, Wolke JCC, Jansen JA (2007) Ceramic composites as matrices and scaffolds for drug delivery in tissue engineering. Advanced Drug Delivery Reviews 59: 234–248.

Hench LL (1991) Bioceramics: from concept to clinic. Journal of the American Ceramic Society 74: 1487–1510.

Hench LL (1998) Bioceramics. Journal of the American Ceramic Society 81: 1705–1728.

Hench LL, Polak JM (2002) Third-generation biomedical materials. Science 295(8): 1014–1017.

Hench LL, West JK (1990) The sol-gel process. Chemical Reviews 90: 33–72.

Hench LL, Wilson J, eds. (1993) An Introduction to Bioceramics, Advanced Series in Ceramics, vol. 1. Singapore: World Scientific.

Hench LL, Splinter RJ, Allen WC, Greenlee TK (1972) Bonding mechanisms at the interface of ceramic prosthetic materials. Journal of Biomedical Materials Research 2: 117–141.

Hubbell JA (1999) Bioactive materials. Current Opinion in Biotechnology 10: 123–129.

ISO (1994) Implants for surgery—Ceramic materials based on high purity alumina. ISO 6474:1994.

Jarcho M (1981) Calcium phosphate ceramics as hard tissue prosthetics. Clinical Orthopaedics and Related Research 157: 259–278.

Kamegai A, Shimamura N, Naitou K, Nagahara K, Kanematsu N, Mori M (1994) Bone formation under the influence of bone morphogenetic protein / self-setting apatite cement composite as delivery system. Bio-medical Materials and Engineering 4: 291–307.

Kamitakahara M, Ohtsuki C, Miyazaki T (2007) Coating of bone-like apatite for development of bioactive materials for bone reconstruction. Biomedical Materials 2: R17–R23.

Kasuga T, Abe Y (1999) Calcium phosphate invert glasses with soda and titania. Journal of Non-Crystalline Solids 243: 70–74.

Kim HM, Himeno T, Kawashita M, Kokubo T, Nakamura T (2004) The mechanism of biomineralization of bone-like apatite on synthetic hydroxyapatite: an in vitro assessment. Journal of the Royal Society, Interface 1: 17–22.

Klawitter JJ, Hulbert SF (1971) Application of porous ceramics for the attachment of load bearing orthopaedic applications. Journal of Biomedical Materials Research 2: 161–229.

Knepper-Nicolai B, Reinstorf A, Hofinger I, Flade K, Wenz R, Pompe W (2002) Influence of osteocalcin and collagen I on the mechanical and biological properties of biocement D. Biomolecular Engineering 19: 227–231.

Kneser U, Schaefer DJ, Munder B, Klemt C, Andree C, Stark GB (2002) Tissue engineering of bone. Minimally Invasive Therapy and Allied Technologies 11(3): 107–116.

Knowles JC, Franks K, Abrahams I (2001) Investigation of the solubility and ion release in the glass system K2O-Na2O-CaO-P2O5. Biomaterials 22: 3091–3096.

Kokubo T (1998) Apatite formation on surfaces of ceramics, metals and polymers in body environment. Acta Materialia 46: 2519–2527.

Komlev VS, Barinov SM, Koplik EV (2002) A method to fabricate porous spherical hydroxyapatite granules intended for time-controlled drug release. Biomaterials 23: 3449–3454.

Krajewski A, Ravaglioli A, Fiori C, Dalla Casa R (1982) The processing of hydroxyapatite rolled sections. Biomaterials 3: 117–120.

Lawson S (1995) Environmental degradation of zirconia ceramics. Journal of the European Ceramics Society 15: 485–502.

LeGeros RZ (1988) Calcium phosphate materials in restorative dentistry: a review. Advances in Dental Research 2(1): 164–180.

LeGeros RZ, Lin S, Rohanizadeh R, Mijares D, LeGeros JP (2003) Bipahsic calcium phosphate bioceramics: preparation, properties and applications. Journal of Materials Science: Materials in Medicine 14: 201–209.

Levitt GE, Crayton PH, Monroe EA, Condrate RA (1969) Forming methods for apatite prosthesis. Journal of Biomedical Materials Research 3: 683–685.

Li R, Clark AE, Hench LL (1991) An investigation of bioactive glass powders by sol-gel processing. Journal of Applied Biomaterials 2: 231–239.

Marouf HA, Quayle AA, Sloan P (1990) In vitro and in vivo studies with collagen/hydroxyapatite implants. The International Journal of Oral and Maxillofacial Implants 5: 148–154.

Marti A (2000) Inert bioceramics (Al$_2$O$_3$, ZrO$_2$) for medical application. Injury 31: D33–D36.

Masonis JL, Bourne RB, Ries MD, Mccalden W, Salehi A, Kelman DC (2004) Zirconia femoral head fractures: a clinical and retrieval analysis. The Journal of Arthroplasty 19(7): 898–905.

Montufar EB, Traykova T, Gil C, Harr I, Almirall A, Aguirre A, Engel E, Planel JA, Ginebra MP (2010) Foamed surfactant solution as a template for self-setting injectable hydroxyapatite scaffolds for bone regeneration. Acta Biomaterialia 6: 876–885.

Murugan R, Ramakrishna S (2005) Development of nanocomposites for bone grafting. Composites Science and Technology 65: 2385–2406.

Nancollas GH, Wu W (2000) Biomineralization mechanism: a kinetics and interfacial energy approach. Journal of Crystal Growth 211: 137–142.

Navarro M, Ginebra MP, Clement J, Martinez S, Avila G, Planell JA (2003a) Physicochemical degradation of titania-stabilized soluble phosphate glasses for medical applications. Journal of the American Ceramic Society 86: 1345–1352.

Navarro M, Ginebra MP, Planell JA (2003b) Cellular response to calcium phosphate glasses with controlled solubility. Journal of Biomedical Materials Research. Part A 67-A: 1009–1015.

Navarro M, Ginebra MP, Planell JA, Zeppetelli S, Ambrosio L (2004a) Development and cell response of a new biodegradable composite scaffold for guided bone regeneration. Journal of Materials Science. Materials in Medicine 15: 419–422.

Navarro M, del Valle S, Martinez S, Zeppetelli S, Ambrosio L, Planell JA, Ginebra MP (2004b) New macroporous calcium phosphate glass ceramic for guided bone regeneration. Biomaterials 25: 4233–4241.

Navarro M, Ginebra MP, Planell JA, Barrias CC, Barbosa MA (2005) In vitro degradation behavior of a novel bioresorbable composite material based on PLA and a soluble CaP glass. Acta Biomaterialia 1: 411–419.

Nogiwa-Valdez AA, Cortés-Hernández DA, Almanza-Robles JM, Chávez-Valdez A (2006) Bioactive zirconia composites. Materials Science Forum 509: 193–198.

Okazaki M, Ohmae H, Takahashi J, Kimura H, Sakuda M (1990) Insolubilized properties of UV-irradiated Co3-apatite-collagen composites. Biomaterials 11: 568–572.

Ooms EM, Wolke JGC, Van de Heuvel MT, Jeschke B, Jansen JA (2003) Histological evaluation of the bone response to calcium phosphate cement implanted in cortical bone. Biomaterials 24: 989–1000.

Otsuka M, Matsuda Y, Suwa Y, Fox J, Higuchi W (1994a) Novel skeletal drug-delivery system using sel-setting calcium phosphate cement: 3. Physicochemical properties and drug-release rate of bovine insulin and bovine albumin. Journal of Pharmaceutical Sciences 83: 255–258.

Otsuka M, Matsuda Y, Suwa Y, Fox J, Higuchi W (1994b) A novel skeletal drug delivery system using a self-setting calcium phsophate cement: 2. Physicochemical properties and drug release of the cement containing indomethacin. Journal of Pharmaceutical Sciences 83: 611–615.

Otsuka M, Matsuda Y, Suwa Y, Fox J, Higuchi W (1994c) A novel skeletal drug delivery system using a self-setting calcium phosphate cement: 5. Drug release behaviour from a heterogeneous drug-loaded cement containing an anticancer drug. Journal of Pharmaceutical Sciences 83: 1565–1568.

Oyane A, Kim HM, Furuya T, Kokubo T, Miyazaki T, Nakamura T (2003) Preparation and assessment of revised simulated body fluids. Journal of Biomedical Materials Research. Part A 65A(2): 188–195.

Perut F, Montufar EB, Ciapetti G, Santin M, Salvage J, Traykova T, Planell JA, Ginebra MP, Baldini N (2011) Novel soybean/gelatine-based bioactive and injectable hydroxyapatite foam: material properties and cell response. Acta Biomaterialia 7: 1780–1787.

Piconi C, Maccauro G (1999) Zirconia as a ceramic biomaterial. Biomaterials 20: 1–25.

Ratner BD, Hoffman AS, Schoen FJ, Lemons JE, eds. (1996) Biomaterials Science: An Introduction to Materials in Medicine. New York: Academic Press.

Ripamonti U, Ma S, Reddi AH (1992) The critical role of geometry of porous hydroxyapatite delivery system in induction of bone by osteogenin, a bone morphogenetic protein. Matrix 12: 202–212.

Sadeghian Z, Heinrich JG, Moztarzadeh F (2005) Preparation of highly concentrated aqueous hydroxyapatite suspensions for slip casting. Journal of Materials Science 40(17): 4619–4623.

Suchanek W, Yoshimura M (1998) Processing and properties of hydroxyapatite-based biomaterials for use as hard tissue replacement implants. Journal of Materials Research 13(1): 94–117.

Sun L, Berndt CC, Gross KA, Kucuk A (2001) Review: material fundamentals and clinical performance of plasma sprayed hydroxyapatite coatings. Journal of Biomedical Materials Research 58(5): 570–592.

Tas AC, Bhaduri S (2004) Rapid coating of Ti6Al4V at room temperature with a calcium phosphate solution similar to 10× simulated body fluid. Journal of Materials Research 19(9): 2742–2749.

TenHuisen KS, Martin RI, Klimkiewicz M, Brown PW (1995) Formation and properties of a synthetic bone composite: hydroxyapatite-collagen. Journal of Biomedical Materials Research 29: 803–810.

Uo M, Mizuno M, Kuboki Y, Makishima A, Watari F (1998) Properties and cytotoxicity of water soluble Na2O-CaO-P2O5 glasses. Biomaterials 19: 2277–2284.

Vaahtio M, Peltola T, Hentunen T, Ylänen H, Areva S, Wolke J, Salonen JI (2006) The properties of biomimetically processed calcium phosphate on bioactive ceramics and their response on bone cells. Journal of Materials Science: Materials in Medicine 17: 1113–1125.

Van Wezel AL (1967) Growth of cell-strains and primary cells on micro-carriers in homogenous culture. Nature 216: 64–65.

Walter G, Vogel J, Hoppe U, Hartmann P (2001) The structure of CaO-Na2O-MgO-P2O5 invert glass. Journal of Non-Crystalline Solids 296: 212–223.

Wise DL, Trantolo DJ, Lewandrowski KU, Gresser JD, Cattaneo MV, eds. (2000) Biomaterials Engineering and Devices: Human Applications, Vol. 2: Orthopaedic, Dental and Bone Graft Applications. Totowa, NJ: Humana Press.

Yu D, Wong J, Matsuda Y, Fox J, Higuchi W, Otsuka M (1992) Self-setting hydroxyapatite cement: a novel skeletal drug-delivery system for antibiotics. Journal of Pharmaceutical Sciences 81: 529–531.

Yuan H, van Blitterswijk CA, de Groot K, de Bruijin JD (2006) Cross-species comparison of extopic bone formation in biphasic calcium phosphate (BCP) and hydroxyapatite (HA) scaffolds. Tissue Engineering 12(6): 1607–1615.

Zhong J, Greenspan DC (2000) Processing and properties of sol-gel bioactive glasses. Journal of Biomedical Materials Research. Part B, Applied Biomaterials 24: 694–701.

7

BIOFUNCTIONAL BIOMATERIALS OF THE FUTURE

Mário Barbosa, Gary Phillips, and Matteo Santin

7.1 CLINICALLY LED NEXT GENERATION BIOMATERIALS

This book has provided general concepts and specific examples of biomimicking, bioresponsive, and bioactive biomaterials. Their different physicochemical characteristics and biocompatibility potential, the principles underlying their design, synthesis, and engineering have all been tightly linked to specific clinical applications. Furthermore, regardless of their differences, a common denominator has clearly emerged: the ability of these materials to participate in and contribute to the regeneration and/or homeostasis process of the specific tissue in which they are implanted. Whatever their intended function, be it the manufacture of medical implants or tissue engineering constructs, the features of these biomaterials have been developed with the aim of favoring their interaction with components of the living cells, both biochemical (e.g., extracellular matrix [ECM] macromolecules) and cellular. This has been a significant paradigm shift that has led to the development of highly performing materials emanating from numerous research projects investigating processes such as cell adhesion and proliferation. Most of these projects have taken into account the relationship between cell behavior and the physicochemical properties of the underlying substrate. More recently, investigations have emerged that try to pinpoint the transduction of mechanical stimuli from the underlying substrates to the cell body [Chen et al. 2004]. Cell proliferation and differentiation as well as expression of particular markers and secretion of cytokines

Biomimetic, Bioresponsive, and Bioactive Materials: An Introduction to Integrating Materials with Tissues, First Edition. Edited by Matteo Santin and Gary Phillips.
© 2012 John Wiley & Sons, Inc. Published 2012 by John Wiley & Sons, Inc.

and growth factors have been studied in depth across a range of different biomaterials and cell types. As a consequence, an enormous body of data has been collected that have led to new insights regarding the various aspects of the so-called interfacial biology [Kirkpatrick et al. 1997].

However, it is felt that scientists are still scratching the surface of what is the complex series of events leading to tissue repair. For example, at the time of writing, there are still too many studies focusing on the adverse reaction caused by the inflammatory response to implants, and few have systematically investigated the ability of macrophages to promote tissue repair. Also, the mechanisms leading to the mobilization of endogenous stem cells from their niche toward the damaged tissue have not yet been fully addressed. Finally, the performance of a given biomaterial when implanted in a pathological rather than a healthy injured tissue has been overlooked.

These considerations would suggest that the next generation of biomaterials will emerge from the wide platform of biomimicking, bioresponsive, and bioactive biomaterials recently designed, but that more focused projects will be needed to address the problem of tissue repair in specific clinical scenarios. In this chapter, examples are given of strategies that may be implemented in the not too distant future to generate highly performing biomaterials.

7.1.1   Wound Dressings and Dermal Substitutes

Despite its complexity, the repair of damaged skin is still pursued with biomaterials with a limited biomimicking and bioactive potential. Wound dressings are primarily hydrogels/hydrocolloids and foams made of synthetic and natural polymers able to partially mimic the skin ECM [Santin 2010]. Little consideration is paid to the design of commercially viable and clinically performing wound dressings able to respond to the different clinical scenarios. A typical example is their widespread use in chronic foot ulcers. Chronic foot ulcers can be classified by various methods mostly focusing on the degree of damage of the tissue [Mostow 1994]. Podiatrists are presented with a wide spectrum of ulcers ranging from those relatively superficial to those so deep that they involve the underlying bony tissues. Regardless of this variegate clinical framework, no particular change in the wound dressing specification is made. Despite the demand for bioactive wound dressings, the currently available products have no ascertained bioactivity and, moreover, they do not address the biochemical and cellular needs of the different tissue layer damaged by the pathology.

Attempts in this direction have rather been made in the design of dermal substitutes which are used to treat burns of different degrees [Santin 2010]. Indeed, most of the dermal substitutes available on the market are made of biopolymers such as collagen and glycosaminoglycan (GAG) that are natural components of the dermis. In recognition of their important role but insufficient performance, dermal substitutes with seeded cells have been made available [Santin 2010]. Although these approaches have led to improved clinical outcomes, scarring and death by infections are still recurrent in patients affected by extended and deep burns. Major issues are the ability of these dermal substitutes to support angiogenesis and to participate in the tissue remodeling process. Plastic surgeons still aim at obtaining the formation of granulation tissue as the only possible alternative to achieve the integration of the dermal substitute. The rate of resorption of the collagen forming the bulk of the dermal substitute is not always tuned with the rate of tissue remodeling. As a consequence of the formation of granulation tissue and the not tuned biomaterial resorption rate, scar often forms. In any case, the repaired skin will still lack hair, hair follicles, and glands which play a fundamental role in the healthy skin because of their role in transpiration and, in the case of the hair follicles, as a reservoir of stem cells. Likewise, in most of the cases, the regeneration of an epithelial layer relies on the ability of the dermal substitute to support epithelial cells, and little consideration has been given for the specific needs of the cuboidal cells with a progenitor character that populate the basal layer of the epidermis.

7.1.2 Vascular Grafts and Cardiovascular Stents

Vascular grafts are devices aiming at the repair of large diameter vessels such as the aorta and the femoropopliteal veins [Ramakrishna et al. 2001]. Despite their key role in saving the life of many patients worldwide, their material and engineering design is relatively basic and leads to limited clinical performances (from 1 to 5 years). Textile materials such as Dacron, expanded poly(tetrafluoroethylene) (ePTFE), and polyurethanes have been used in combination with a metal stent to support the in-growth of fibrotic tissue capable of bypassing the main aneurysm or vessel occlusion. Although the texture and surface properties of these materials are able to encourage tissue in-growth and implant stabilization, these devices are destined to fail because of thrombogenicity and graft occlusion [Ramakrishna et al. 2001]. This is not surprising if their chemical and structural properties are compared with the sophisticated features of a natural blood vessel. The formation of a neointimal tissue based on a monolayer of endothelial cells as well as the multilayered and highly oriented deposition of the blood vessel media populated by smooth muscle cells and elastin fibers are not controlled by these devices, but rather compromised by their presence. As a consequence, the anti-thrombogenic character of the natural intima and the contractile properties of the media are irreversibly compromised. Likewise, cardiovascular stents consist of a tubular metal mesh deployed to eliminate occlusions (stenosis) of small diameter blood vessels such as the coronary arteries, the lumen patency of which is affected by the thickening of an atherosclerotic plaque [Santin et al. 2005]. Although these devices can be inserted by a minimally invasive procedure through catheter insertion, their life span is also limited. In fact, as a consequence of the stent balloon-driven expansion, their implantation causes the disruption of the endothelial layer paving the blood vessel, and their foreign surface later induces a host response [Santin et al. 2005]. This host response leads to an uncontrolled neointimal tissue formation that is characterized by a thick layer of smooth muscle cells causing restenosis, the recurrent occlusion of the vessel lumen. The metal mesh is merely a mechanical support to the atherosclerotic vessel and has no biomimetic architecture. The coating of these metal devices with biomimetic polymers such as those presented in Chapter 4 aims to mimic the host response, but it is not capable of encouraging the formation of a physiological tissue. As for vascular grafts, the architecture of the stented vessel is significantly altered by the presence of the device, and clinical failure is a frequent outcome.

It is evident that, while the currently available devices have been conceived by taking into account the mechanical requirements, little effort has been paid to mimic the histological features of the blood vessels. Future devices for blood vessel treatment will surely benefit from a design that is inspired by the biochemical and structural properties of the blood vessel wall.

7.1.3   Joint Implants and Cartilage Tissue Engineering

Arthritis and osteoarthritis are among the most debilitating pathologies in modern society. The aging population, as well as the increasing number of individuals undergoing regular and sometimes intensive sport training, are prone to these pathologies affecting joints, especially those of the lower limbs. Like skin, articular cartilage is a tissue with a very sophisticated structure that has evolved to play a role as a fine shock absorber during ambulation. In a large number of cases, the development of arthritis and osteoarthritis compromises the integrity of this tissue that is the only tissue that is unable to undergo spontaneous repair. Cartilage has been one of the first tissues in which repair has been attempted by regenerative medicine approaches based on the implantation of autograft tissue (i.e., mosaicplasty) and autologous cells (autologous chondrocyte implantation [ACI]) [Oakes 2009]. However, both mosaicplasty and ACI have a very limited performance that is due to the lack of integration of the implanted autologous material into the damaged tissue. Biomaterials such as hyaluronan and fibrin glue have been employed in attempts to support endogenous or transplanted cells in tissue repair [Jiang et al. 2007; Oakes 2009]. However, so far, no satisfactory clinical outcome has been achieved. This clinical failure is underpinned by a lack of adequate microenvironments for cell survival and activity. In a healthy cartilage, chondrocytes are encapsulated in the so-called chondron, a structural and functional unit where the cells' immediate surrounding is characterized by the presence of collagen type II and by its ability to bind and gradually release growth factors such as the insulin-like growth factor 1 (IGF-1) and the transforming growth factor beta 1 (TGF-β1) [Oakes 2009]. At biomacromolecular level, cartilage is composed not only of GAGs and proteoglycans (PGNs), but also of a finely oriented architecture of collagen fibers. The combination of these features confers to the tissue the biomechanical properties of a shock absorber and at the cellular level the appropriate haemostasis. No biomaterial has ever been produced that tries to mimic all (or most) of these features.

As a consequence, degeneration of the cartilage leads to painful and very disabling conditions and ultimately to the surgical implantation of artificial joints including knee and hip prosthesis. Here, the need to replace the damaged joint cartilage (a tissue of few millimeters in thickness) massively compromises the limb structure. Indeed, the fixing of these implants in the bone marrow canal leads to a damage of the bony tissue. The metallic and ceramic materials described in Chapters 5 and 6 have been an attempt to mimic the bony tissue properties and, therefore, to encourage the bone repair. However, none of these materials for load-bearing applications has been able to achieve a complete repair of bone and to sustain its later remodeling.

Until a successful regeneration of the cartilage will be achieved, it is inconceivable to think about clinical treatments not relying on these load-bearing metal-based implants, which often include a high molecular weight polyethylene (HMWPE) component to minimize friction. However, the release of wear particles, mainly from the polymer, is the cause of aseptic loosening, which is a major problem associated with these implants. However, the concepts and technologies for biomimetic surface functionalization described in Chapters 1 and 4 could also be applied to the functionalization of metals with biomacromolecules able to induce full osteointegration of the prosthesis in parts not subjected to wear.

7.1.4   Bone Fillers

Bone repair and remodeling has been more successfully obtained with the use of bone fillers, some of them with clear biomimetic (i.e., ceramics) and bioactive (i.e., bioglasses) properties (see Chapter 6). Biopolymers (e.g., collagen) have also been used to fill bone defects, but their tissue repair potential is questionable. Composite bone fillers have also been investigated as bone fillers. Polylactidades and fumarates are among the polymers used. Again, none of these materials has been engineered to mimic all the main components of bone.

7.1.5   Nerve Guides

The simplistic approach that currently accompanies the manufacture of nerve guides for peripheral nerve repair somehow parallels those adopted for the production of vascular grafts. Nerve guides are simply tubular materials made of synthetic or natural, nondegradable or biodegradable polymers [Bender et al. 2004]. These are merely physical guides aimed at directing the axonal outgrowth by neural cells. No consideration has been so far paid to the manufacture of neural guides able to promote the proliferation of the cell type responsible for the axonal outgrowth, the glial cells. A vaguely biomimetic approach has so far been adopted at a research level whereby some guides have been designed with microgrooves able to direct axonal outgrowth [Potucek et al. 2009]. This approach does not take into account that a healthy nerve axon is encased in a connective tissue, the perineurium, as well as blood vessels necessary for tissue survival, and a similar vascularized collagenic matrix embeds the axonal bundles. Even those neural guides more closely resembling nature (i.e., collagen-based neural guides) cannot discriminate between the adhesion and proliferation of inflammatory cells, fibroblasts, glial cells, neural cells, and endothelial cells; the latter three are known to be supported by specific proteins such as laminin rather than by collagen.

7.1.6   Ophthalmologic Devices

The delicate balance of tissue compartments characterizing our eyes also suffers from simplistic technological approaches. Devices such as contact lenses and intraocular lenses are based on hydrogel technology that has very little ability to mimic the original tissue. Bioinert biomaterials rather than bioresponsive polymers have been developed in the last few decades and although they are able to control host response, they are not able to limit pathological angiogenesis (see Chapter 4).

7.2 BIOMACROMOLECULE-INSPIRED BIOMATERIALS

Although many research papers, patents, and products have been labeled under the category of biomimicking biomaterials, the level of bioinspiration is rather limited and, sometimes, arguable. As a consequence, very few examples of work at both research and industrial level stand out as truly biomimicking. Even fewer emerge as bioactive and bioresponsive. This section provides some examples of research activities that have reacted to the need for biomaterials that are capable of both responding to and interacting with the biological environment of the tissue to be repaired. In particular, biomaterials that have been inspired by features of the biomacromolecules of the ECM and that have the potential to provide solutions to the neglected areas identified in Sections 7.1.1–7.1.6 will be presented.

7.2.1 Artificial Laminin

The basement membrane is the thin protein layer on which cells such as keratinocytes grow to form epidermis [Inoue et al. 2005]. Basement membranes also support the growth of neurons, vascular endothelial cells, and hepatocytes [Kowtha et al. 1998; Freire et al. 2002]. BD Matrigel™ is a commercial product based on extracted collagen type IV (31%), laminin (56%), and entactin (8%) that has been proven to be particularly efficacious in the culturing of endothelial cells and, as a consequence, the substrate of choice for the study of angiogenesis in vitro [Finkenzeller et al. 2009]. Although its use in cell culture systems in vitro has been proven successful, its use in vivo is not advised because of both the tumorogenic potential, possibly due to the presence of growth factors and the risk of transmittable diseases from the animal donor. The need for highly reproducible and safe products in regenerative medicine has therefore prompted studies of biomaterials able to mimic the basement membrane and its functions [Lakshmanan & Dhathathreyan 2006]. Thus far, these studies have been focusing on the synthesis of novel biomaterials bearing specific amino acid sequences of laminin. The most known and exploited laminin-mimicking peptide is the -YIGSR- sequence. However, other sequences have been synthesized such as -AGTFALRGDNPQG- and -RKRLQVQLSIRT. These sequences have been exploited for the synthesis of novel biomaterials by grafting the synthetic peptide on the surface of conventional natural biomaterials such as chitosan [Mochizuki et al. 2003]. Studies have also been dedicated to the importance of the synthetic bioligand on solid surfaces. These studies stem from the understanding that the basement membrane role as cell substrate is also determined by its macromolecular assembly that leads to a very well-defined ultrastructure [Sant'Ana Barroso et al. 2008]. Self-assembly of laminin at different pHs has been shown to significantly alter neurite formation and that cultured keratinocytes can produce a highly structured protein-based thin sheet-like network located at the medium/air interface [Sant'Ana Barroso et al. 2008]. The use of such a basement membrane-like material has been suggested for wound healing applications. The limited amount of materials that can be produced through cultures is a limiting factor, and in the future, completely synthetic biomaterials that are capable of resembling both the bioligands and the structure of the basement membrane will be required.

7.2.2 Artificial Elastin

The ECM of blood vessels is mainly composed of collagen, GAGs, and elastin. The latter is a protein that contributes to the elasticity of the tissue and, as a consequence, to its functionality [Santin 2010]. In attempts to mimic this important component, elastin-like polypeptides have been engineered that include the primary sequence of mammalian elastin (-GVGVP-) [Meyer et al. 2001]. In addition to their ability to present an amino acid sequence recognized by cell receptors, these polypeptides are also thermoresponsive as they are able to undergo a phase transition and become insoluble at body temperature. Although the elastin-like polymers have so far been proposed as therapeutic agents in cancer therapy [Bidwell & Raucher 2005], their potential use as biomimetic biomaterials in regenerative medicine applications (e.g., cartilage repair and nonthrombogenic coatings) has been gaining the interest of material scientists. To this end, in a very elegant approach, the elastin-like polypeptide technology has been brought to a three-dimensional (3D) level by stratified ultrathin polyelectrolyte layer-by-layer assembly [Swierczewska et al. 2008]. These ultrathin layers have been alternated with cell layers to obtain tissue engineering constructs. Together with the laminin-mimicking biomaterials, these assemblies could play an important role in controlling angiogenesis in many tissue regeneration applications.

7.2.3 Artificial Collagen

As one of the main components of the ECM of many tissues and its important role in tissue homeostasis, collagen is among the most used natural biomaterials for the repair of both soft and bony tissues. The problem associated with the use of collagen is the risk of transmittable diseases from the donor as well as a host response toward the protein. The first problem has been tackled by the use of recombinant human collagen that, however, is relatively expensive to produce and is still potentially immunogenic. For these reasons, artificial collagen has been proposed that is based on the well-known collagen sequence, Pro-Hyp-Gly, grafted to a synthetic polymer such as poly(ethylene glycol) [Wang et al. 2008]. These conjugates were shown to be able to reversibly bind natural collagen and to modulate the cell adhesion process (i.e., temporary cell repulsion). These tunable features could be exploited for fabricating multifunctional cell substrates and scaffolds for bioresponsive implant coatings and tissue engineering constructs.

Likewise, poly(Xaa-Yaa-Gly)$_n$ peptide polymers have been prepared that can reach a molecular weight of 1×10^6 Da [Paramonov et al. 2006]. These synthetic polymers can rearrange in the form of a collagen-like triple helix that leads to the formation of micrometer-long nanofibers. The mild conditions of their synthesis offer the opportunity to include amino acid with side chains leading to specific biofunctionalities. For example, phosphoserine could be introduced to favor the biomineralization of these polymers in applications where bone tissue repair is needed.

7.2.4 GAG- and PGN-Mimicking Biomaterials

GAGs and PGNs are highly hydrophilic molecules playing an important role in the water storage and shock absorbing properties of articular cartilage. As for all biopolymers, whether from tissue or of recombinant origin, GAG and PGN suffer from both batch-to-batch reproducibility and risks of transmittable disease. GAG-mimicking biomaterials have therefore been designed. Various water-soluble glycopolymers with high saccharide contents have been obtained by the radical polymerization of cyanoxyl-persistent radicals [Grande et al. 2001]. The advantage of these radicals is their ability to act as chain-growth moderators of acrylamide polymerization. During this polymerization process, either nonsulfated or sulfated N-acetyl-d-glucosamine-carrying alkene- and acrylate-derivatized unprotected glycomonomers can be introduced. Monosaccharide-containing homopolymers can also be prepared. Santin et al. have demonstrated the regiospecific synthesis of galactosylated acrylic monomers by thermophilic bacterium enzyme catalysis [Santin et al. 1995]. These monomers can be integrated into the vinyl radical polymerization of hydrogels such as the poly(2-hydroxyethyl methacrylate) polymers to insert a galactose unit recognized by specific cell types such as hepatocytes. Octopus oligosaccharides have also been synthesized that are based on the principle of dendrimeric polymers [Dubber & Lindhorst 2001].

The potential application of these polymers in regenerative medicine has not been yet demonstrated.

7.3 NANOSTRUCTURED BIOMIMETIC, BIORESPONSIVE, AND BIOACTIVE BIOMATERIALS

Section 1.3.1 has shown that tissue biology is ultimately governed by biointeractions taking place at the subcellular level. For example, the integrin-mediated biorecognition processes involved in cell adhesion, which leads to phenomena such as cell migration, proliferation, and differentiation, are established in environments with dimensions ranging within the 0.1 μm scale. Authors have determined the optimal distance between integrin-specific bioligands and found it confined to nanoscale levels. Similarly, the surface of the basement membrane, the collagen IV-based layer where endothelial cells sit, and of other collagenic ECM are characterized by relatively ordered structures from which a nanoscale roughness emerges [Miller et al. 2004]. More importantly, several investigations have proven that cell adhesion is favored by this level of roughness [Sato & Webster 2004]. Similarly, immature bone is characterized by a mineral phase with a grain average size ranging between 10 and 50 nm, while in mature bone this is restricted to 20–50 nm [Ward & Webster 2006]. It has been demonstrated that the characteristics of the osteoblast-deposited mineral phase change when cells are seeded on metallic implants with varying nanotopography [Sandrini et al. 2005].

7.3.1 Nanofabrication of Biomaterials

In the last decade, the goals of biorecognition in biomedical implants and tissue engineering has been pursued by developing biomaterials with structures controlled at the nanoscale (<100 nm) [Miller et al. 2004]. Several synthetic and engineering methods have been adopted to achieve this goal at both two-dimensional (2D) and 3D levels [Miller et al. 2004].

7.3.1.1 2D Techniques. One of the techniques used to control the 2D morphology of solid surfaces is lithography. Through this technique, processes can be repeated in a very accurate and relatively nonexpensive manner. A typical lithography procedure can be summarized in three different steps [Miller et al. 2004]:

1. Coating of the substrate with resists that are resistant to irradiation
2. Exposure of the resist to beams of light, electrons, or ions
3. Development of the resist image on the substrate.

In a modification to this method (colloidal lithography), dispersed nanoparticles can be electrostatically bound to the substrate and used as a template to etch it and ultimately to expose a surface with nanotopographical features (e.g., nanocolumns).

Chemical and physical methods include the etching of polymeric and metal surfaces with alkali or with highly reactive species such as oxygen radicals, as well as the deposition of thin films of ceramics (see Chapter 5). Sophisticated material casting methods and molecular imprinting have also been used to mimic the basement membrane of the endothelium or to favor the adsorption of specific proteins in their native conformation [Shi et al. 1999].

7.3.1.2 3D Techniques. Polymer scaffolds have been produced by techniques such as electrospinning and phase separation. Electrospinning involves the generation of a relatively large electric potential (15 to 30 kV) between a polymer solution and a target (Figure 7.1) [Miller et al. 2004]. Once the electric field is established, the electrostatic forces stretch the dropping polymer solution into a cone, the so called Taylor

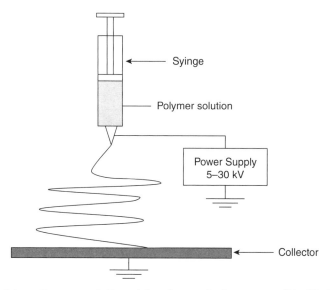

Figure 7.1. Schematic representation of the electrospinning process. (Modified from Miller et al. 2004.)

cone. When critical values are reached the electrostatic forces overtake the surface tension of the polymer solution and form a fluid jet at the tip of the cone. Traveling towards the collector the fluid jet starts to whip around while condensing by solvent evaporation. In a typical electrospinning process, the jet reaches the collector assuming looping paths which leads to the formation of nonwoven mats with fibers aligned at the nanoscale level. Adjustment of the experimental conditions can allow the formation of finely aligned fibres. Although the technique has been known for over 70 years interest in its application to tissue engineering has increased exponentially in the last decade. The ability to electrospun a wide range of polymers to produce nanofibres that mimic the fibrous components of the ECM has been a major reason for its extensive use. A comprehensive review on the technique and materials employed is provided by Pham et al. [2006].

Nanostructuring through phase separation can be obtained by treating a polymer solution with an immiscible solvent or by cooling the solution by well-tuned processes including freeze-drying [Miller et al. 2004]. In thermally induced phase separation (TIPS) methods, the temperature of the solution is lowered to produce phase separation. Like electrospinning, phase separation methods can be used to obtain scaffolds consisting of nanofibers, although its major use has been in the production of porous 3D scaffolds. Freeze-gelation has been proposed as an alternative method to freeze-drying [Ho et al. 2004]. In this method, the frozen polymer solution is immersed in a gelation environment at a temperature below the freezing point of the polymer solution, thus enabling the formation of a gel before the drying stage.

7.3.1.3 Polymeric Dendrimers.
Dendrimers (from the old Greek "dendron," which means "tree") are hyperbranched polymers which can be obtained by liquid or solid phase synthesis [Esfand & Tomalia 2001; Tomalia 2005]. These polymers can be obtained by different types of reaction. One of the most common synthetic methods is the Michael's addition reaction by which branching of the growing polymer is obtained by adding a methacrylate to a core molecule (Figure 7.2). Next, the branch length is extended by adding a diamine. This method produces poly(amidoamine) (PAMAM) dendrimers of different generations (G_x) from G_0 to G_9. When synthesized by liquid phase reaction, branching occurs in all directions, and spherical dendrimers are obtained. However, this method requires laborious purification of the product. Conversely, when performed in solid phase (e.g., on a peptide matrix support), semi-dendrimers are obtained (Figure 7.3). The advantage of this method is that a pure product can be acquired by simple washing of the resin before cleaving the synthesized molecules off from the solid support.

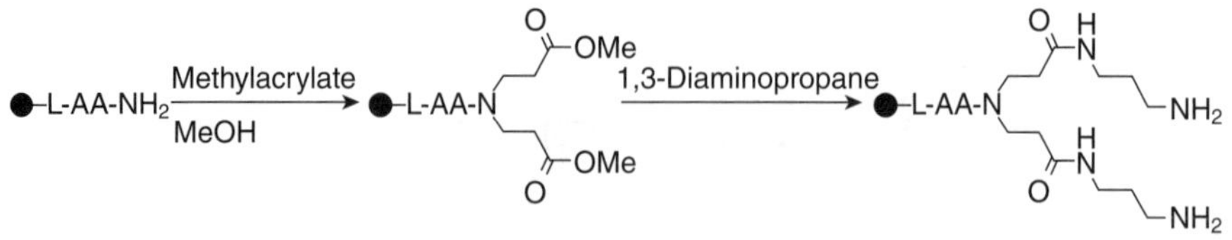

Figure 7.2. Solid phase synthesis reaction of PAMAM dendrimer.

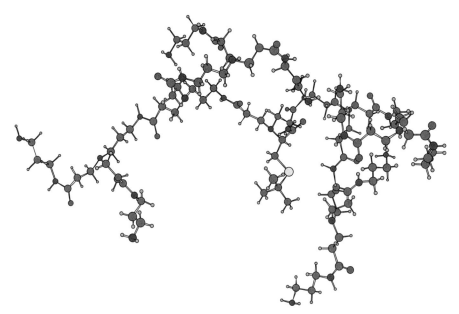

Figure 7.3. Computer modeling of a typical PAMAM G3 semi-dendromer. (By Dr. Peter Cragg, University of Brighton, UK).

Different types of dendrimers have been synthesized by using peptide- (e.g., poly-lysine dendrimers) or oligosaccharide-based synthesis [Dubber & Lindhorst 2001]. Peptide synthesis has also been used to functionalize dendrimers with peptidic bioli-gands for cell integrins [Monaghan et al. 2001]. Thus far, given their ability to bind DNA through electrostatic interactions, dendrimers have been proposed mainly as car-riers for gene delivery [Fischer et al. 2003] and commercially used as carrier for in vitro cell transfection.

7.3.1.4 Self-Assembling Peptides. Self-assembly of biological molecules is present in numerous natural tissues. Weak physical bonds, namely hydrogen bonds, electrostatic interactions, hydrophobic interactions, and van der Waals forces, govern it. The cell membrane is a typical example of self-assembly in nature. Complex and strong structures can be formed by self-assembly, in spite of the weakness of individual bonds. The ability of biological systems to adapt to changes in the environment without rupturing stronger bonds is related to the reversible nature of self-assembly. Nanostruc-tures have also been produced through the synthesis of self-assembling peptides. This method can be considered as the "true" biomimetic method as both the self-assembling and the exposure of bioligands mimic processes adopted by nature to build up ECM. Indeed, self-assembly is the process by which specific functional groups in synthetic or natural polymers drive interactions with other polymeric chains to form spontane-ously ordered structures. This molecular organization leads to supramolecular structures and nanofibers [Hosseinkhani et al. 2006]. For example, amphiphilic peptides (peptide sequences formed by amino acids carrying both hydrophobic and hydrophilic domains) have been synthesized that are capable of forming nanofibers of a diameter smaller than 10 nm [Paramonov et al. 2006]. Similarly, interactions and supramolecular structuring has been obtained by the combination of functional groups capable of responding to pH variations [Jang et al. 2004]. Recently, peptides assembling into nanofibers have been tested as carriers for mesenchymal stem cells showing that they can induce their differentiation [Hosseinkhani et al. 2006]. Likewise dendrimers, branched peptide amphiphile molecules bearing integrin-specific bioligands (e.g., RGDs), have been synthesized [Guler et al. 2006]. The self-assembling of these molecules leads to cylindrical nanofibers, and the branched nanoarchitecture makes the bioligands more accessible to the cells while increasing their surface density. Although this type of biomaterials has received much attention in recent years, the ability of peptides to self-assemble into beta-sheet structures was already reported in 1993 by Zhang et al. After the discovery of a simple repetitive sequence—AEAEAKAKAEAEAKAK—found in a yeast protein called zuotin, he has developed a series of self-assembling peptides for a variety of applications, including neural regeneration, encapsulation of chondrocytes, osteoblast differentiation, and in vitro culture of hepatocytes [Zhao & Zhang 2007].

7.4   CONCLUSIONS

This book has provided an overview on the latest generation of biomaterials and their future perspectives. The various examples of biomimetic, bioresponsive, and bioactive biomaterials that have been presented clearly demonstrate that a significant improvement of the clinical treatment by medical implants and tissue engineering construct can be obtained only by materials inspired by nature. Early interactions between implants and tissues have been greatly enhanced by this approach as it has led to medical implants that are able to integrate better both with the host tissue and its dynamic homeostasis. Monolith materials and coatings are now available that can drive various biochemical and cellular components of a tissue during its phases of repair. Different levels of biomimicking can now be implemented in industrially viable products that include either bioinspired structures or bioactivity. Furthermore, the advent of nanostructured materials has added another level to the biomimicking strategies. These materials now allow the bottom-up assembly of structures capable of either bearing a multifunctional character or of acquiring a completely new function. These assembling biocompetent nanostructures appear to be among the best solutions to fulfill the needs of minimally invasive treatments (e.g., by injection). The number of biomimicking implants that have already been made available to clinical practice, together with the exponential number of pioneering research projects on bioinspired biomaterials, represents a flexible platform on which future highly performing treatments will be based. Scientists, industrialists, and clinicians are called to look at this new age with much enthusiasm and determination to provide new knowledge and products. It is envisaged that easy-to-handle products will be made available in the form of bioactive cartilage and nerve nanoglues, pathology-tailored synthetic morphogens, regenerating patches for myocardium infarct healing, artificial stem cell niches, and self-structuring histodomains. These products will be able to jump-start tissue regeneration processes that have been compromised by trauma or disease by tuning the behavior of endogenous and transplanted cells. They will gradually lead us from the concept of tissue replacement and implant integration to tissue and organ regeneration. The concerted actions of the different stakeholders will be required to make sure that these technologies are made available while paying full respect to the patients' interest and safety.

EXERCISES/QUESTIONS FOR CHAPTER 7

1. Which are the main clinical applications in which wound dressings are used? List the ideal features of a wound dressing and their current limitations in the given clinical examples.
2. Describe the rationale underpinning the use of nerve guides in peripheral nerve repair and provide examples of typical clinical applications where these devices are used.
3. Elastin-like and collagen-like peptides have been synthesized to produce artificial extracellular matrix components. Critically discuss their potential in clinical applications by taking into account their bioactivity and commercial feasibility.
4. What is the role of glycosaminoglycans and proteoglycans in tissue homeostasis and repair?
5. Provide examples of synthetic materials able to mimic the glycosaminoglycans and proteoglycans of the extracellular matrix.
6. Provide at least two examples of nanostructured biomaterials and highlight their advantage in regenerative medicine applications.
7. What are the main lithography procedures adopted to control the 2D topography of biomaterial surfaces?
8. Describe engineering techniques that can be implemented for the construction of 3D scaffolds for tissue engineering.
9. Give a definition of dendrimeric materials, describe a typical method to synthesize them, and provide examples of their application in cell and/or tissue engineering.
10. What is the most known amino acid sequence leading to the self-assembling of peptides?

REFERENCES

Bender MD, Bennett JM, Waddell RL, Doctor JS, Kacey G, Marra KG (2004) Multi-channeled biodegradable polymer/CultiSpher composite nerve guides. Biomaterials 25(7–8): 1269–1278.

Bidwell GL III, Raucher D (2005) Application of thermally responsive polypeptides directed against c-Myc transcriptional function for cancer therapy. Mol Cancer Ther 4(7): 1076–1085.

Chen CS, Tan J, Tien J (2004) Mechanotransduction at cell-matrix and cell-cell contacts. Annu Rev Biomed Eng 6: 275–302.

Dubber M, Lindhorst TK (2001) Threalose-based octopus glycosides for the synthesis of carbohydrate-centered PAMAM dendrimers and thiourea-bridged gycoclusters. Org Lett 3: 4019–4022.

Esfand R, Tomalia DA (2001) Poly(amidoamine) (PAMAM) dendrimers: from biomimicry to drug delivery and biomedical applications. Drug Deliv Today 6: 427–436.

Finkenzeller G, Graner S, Kirkpatrick CJ, Fuchs S, Stark GB (2009) Impaired in vivo vasculo-genic potential of endothelial progenitor cells in comparison to human umbilical vein endo-thelial cells in a spheroid-based implantation model. Cell Prolif 42: 498–505.

Fischer D, Li Y, Ahlemeyer B, Krieglstein J, Kissel T (2003) In vitro cytotoxicity testing of polycations: influence of polymer structure on cell viability and hemolysis. Biomaterials 24: 1121–1131.

Freire E, Gomes FCA, Linden R, Moura Neto V, Coelho-Sampaio T (2002) Structure of laminin substrate modulates cellular signalling for neuritogenesis. J Cell Sci 115: 4867–4876.

Grande D, Baskaran S, Chaikof E (2001) Glycosaminoglycan mimetic biomaterials: 2. Alkene- and acrylate-derivitized glycopolymers via cyanoxyl-mediated free radical polymerization. Macromolecules 34(6): 1640–1646.

Guler MO, Hsu L, Soukasene S, Harrington DA, Hulvat JF, Stupp SI (2006) Presentation of RGDS epitopes on self-assembled nanofibers of branched peptide amphiphiles. Biomacro-molecules 7: 1855–1863.

Ho MH, Kuo P-Y, Hsieh H-J, Hsien T-Y, Hou L-T, Lai J-Y, Wang DM (2004) Preparation of porous scaffolds by using freeze-extraction and freeze-gelation methods. Biomaterials 25(1): 129–138.

Hosseinkhani H, Hosseinkhani M, Kobayashi H (2006) Design of tissue-engineered nanoscaffold through self-assembly of peptide amphiphile. J Bioact Compat Polym 21: 277–296.

Inoue S, Reinisch C, Tschachler E, Eckhart L (2005) Ultrastructural characterization of an arti-ficial basement membrane produced by cultured keratinocytes. J Biomed Mater Res 73A: 158–164.

Jang JH, Houchin TL, Shea LD (2004) Gene delivery from polymer scaffolds for tissue engineer-ing. Expert Rev Med Devices 1(1): 127–138.

Jiang D, Liang J, Noble PW (2007) Hyaluronan in tissue injury and repair. Annu Rev Cell Dev Biol 23: 435–461.

Kirkpatrick CJ, Wagner M, Kohler H, Bittinger F, Otto M, Klein CL (1997) The cell and molecu-lar biological approach to biomaterial research: a perspective. J Mater Sci Mater Med 8: 131–141.

Kowtha VC, Bryant HJ, Pancrazio JJ, Stenger DA (1998) Influence of extracellular matrix pro-teins on a membrane potentials and excitability in NG108-15 cells. Neurosci Lett 246: 9–12.

Lakshmanan M, Dhathathreyan A (2006) Amphiphilic laminin peptides at air/water interface: effect of single amino acid mutations on surface properties. J Colloid Interface Sci 302(1): 95–102.

Meyer DE, Kong GA, Dewhirst MW, Zalutsky MR, Chilkoti A (2001) Targeting genetically engineered elastin-like polypeptide to solid tumors by local hyperthermia. Cancer Res 61: 1548–1554.

Miller DC, Webster TJ, Haberstroh KM (2004) Technological advances in nanoscale biomaterials: the future of synthetic vascular graft design. Expert Rev Med Devices 1: 259–268.

Mochizuki M, Kadoya Y, Wakabayashi Y, Kato K, Okazaki I, Yamada M, Sato T, Sakairi N, Nomizu M (2003) Laminin-1 peptide-conjugated chitosan membranes as a novel approach for cell engineering. FASEB J 17: 875–877.

Monaghan S, Griffith-Johnson D, Matthews I, Bradley M (2001) Solid-phase synthesis of peptide-dendrimer conjugates for an investigation of integrin binding. ARKIVOC Issue in honor of Prof. Kjell Undheim: 46–53.

Mostow EN (1994) Diagnosis and classification of chronic wounds. Clin Dermatol 12(1): 3–9.

Oakes BW (2009) Basic science and clinical strategies for articular cartilage regeneration/repair. In: Strategies in Regenerative Medicine, ed. M Santin, 395–430. New York: Springer Sciences Business Media LLC 13.

Paramonov SE, Jun HW, Hartgerink JD (2006) Self-assembly of peptide-amphiphile nanofibers: the roles of hydrogen bonding and amphiphilic packing. J Am Chem Soc 128: 7291–7298.

Pham QP, Sharma U, Mikos AG (2006) Electrospinning of polymeric nanofibers for tissue engineering applications: a review. Tissue Eng 12(5): 1197–1211.

Potucek RK, Kemp SWP, Syed NI, Midha R (2009) Peripheral nerve injury, repair and regeneration. In: Strategies in Regenerative Medicine, ed. M Santin, 321–340. New York: Springer Sciences Business Media LLC 10.

Ramakrishna S, Mayer J, Wintermantel E, Leong KW (2001) Biomedical applications of polymer-composite materials: a review. Composites Sci Technol 61: 1189–1224.

Sandrini E, Morris C, Chiesa R, Cigada A, Santin M (2005) In vitro assessment of the osteointegrative potential of a novel multiphase anodic spark deposition coating for orthopaedic and dental implants. J Biomed Mater Res B Appl Biomater 73B: 392–399.

Sant'Ana Barroso MM, Freire E, Limaverde GSCS, Rocha GM, Batista EJO, Wiessmuller G, Andrade LR, Coelho-Sampaio T (2008) Artificial laminin polymers assembled in acidic pH mimic basement membrane organisation. J Biol Chem 283: 11714–11720.

Santin M (2010) Soft tissue applications of biocomposites. In: Biomedical Composites, ed. L Ambrosio, London: Woodhead Publishing.

Santin M, Ross F, Sada A, Peluso G, Improta R, Trincone A (1996) Enzymatic synthesis of 2-beta-D galactopyranosylox ethyl methacrylate (GalEMA) by thermophilic archeon Sulfolobus sulfataricus. Biotechnol Bioeng 49: 217–222.

Santin M, Colombo P, Bruschi G (2005) Interfacial biology of in-stent restenosis. Expert Rev Med Devices 2(4): 429–443.

Sato M, Webster TJ (2004) Nanobiotechnology: implications for the future of nanotechnology in orthopedic applications. Expert Rev Med Devices 1: 105–114.

Shi HQ, Tsai WB, Garrison MD, Ferrari S, Ratner BD (1999) Template-imprinted nanostructured surfaces for protein recognition. Nature 398: 593–597.

Swierczewska M, Hajicharalambous CS, Janorkar AV, Megeed Z, Yarmush ML, Rajagopalan P (2008) Cellular response to nanoscale elastin-like polypeptide polyelectrolyte multilayers. Acta Biomater 4: 827–837.

Tomalia DA (2005) The dendritic state. Materials Today March Issue, 34–46.

Wang AY, Foss CA, Leong S, Mo X, Pomper MG, Yu SM (2008) Spatio-temporal modification of collagen scaffolds mediated by triple helical propensity. Biomacromolecules 9(7): 1755–1763.

Ward BC, Webster TJ (2006) The effect of nanotopography on calcium and phosphorus deposition on metallic materials in vitro. Biomaterials 27: 3064–3074.

Zhang SG, Holmes T, Lockshin C, Rich A (1993) Spontaneous assembly of a self-complementary oligopeptide to form a stable macroscopic membrane. Proc Natl Acad Sci U S A 90(8): 3334–3338.

Zhao X, Zhang S (2007) Designer self-assembling peptide materials. Macromol Biosci 7(1): 13–22.

INDEX

Biomimetic, Bioresponsive, and Bioactive Materials: An Introduction to Integrating Materials with Tissues, First Edition. Edited by Matteo Santin and Gary Phillips.
© 2012 John Wiley & Sons, Inc. Published 2012 by John Wiley & Sons, Inc.

Fibrin
  silk fibroin and, 27
  in tissue regeneration, 8, 14, 15
Fibrin-based biomaterials, 7, 8
Fibrin clots, 7, 8
Fibrin glue
  as an adhesive material, 9
  extracellular matrix and, 6–9
Fibrin networks, 7, 8
Fibrinogen, 7
  in foreign body response, 70
  in tissue regeneration, 14, 15
Fibrinolytic processes, 7
Fibroblasts, 41, 44
  in chronic inflammation, 72
  collagen secretion by, 61
  in connective tissue proper, 59
  in fibrous encapsulation, 74
  in foreign body reaction, 74
  granulation tissue and, 73
  in mucous connective tissue, 58
Fibrocartilage, 65–66, 67
Fibrogenic agents, in fibrous encapsulation, 74
Fibroin, in silk, 26–27
Fibronectin, in cell adhesion, 18
Fibroplasia, as wound healing response, 45
Fibrosis, as response to injury, 68, 69
Fibrotic capsules, 4–5
  porous-metal implants and, 11–12
Fibrotic tissue, 9
Fibrous capsule development (encapsulation)
  PC polymers and, 107
  as response to injury, 68, 69, 74–76
Fibrous layer, in biostable ceramics, 162–163
Films. See also Biofilms; Passivation film
  as anodic oxidation coatings, 152–153, 153–154, 154–157
  coating of metals with, 150
Filtration systems, PC-containing polymers in, 115–116
Finishing, of metals, 150
Firing, in high temperature calcium orthophosphate processing, 175, 176
First-generation biomaterials, 1–5
First-generation ceramics, 161–162
Fluorides, in bioactive glasses, 168, 169
Fluoroplastic tympanostomy tubes, 113

Foaming agents, in calcium phosphate cement formation, 177
Focal adhesion, 16
Focal adhesion kinase (FOK), in cell adhesion, 17–18
Focal complexes, in cell adhesion, 16
Ford, Henry, 28
Foreign body giant cells (FBGCs), in response to injury, 68, 69, 70, 72–73, 73–74
Foreign body reaction, as response to injury, 68, 69, 73–74
Foreign body response, 36, 68–76. See also Immune response
  as response to injury, 68, 69, 70, 71
Fouling, in biosensor systems, 113–114
Free oxygens, in glasses, 167
Freeze-drying, 200
Freeze-gelation, 200
Fretting corrosion
  of biomedical metals, 144, 145
  titanium and titanium alloy resistance to, 149
Frustrated phagocytosis, 69–70, 71, 72, 73
Future, biofunctional biomaterials of, 191–206

GAG-mimicking biomaterials, 197–198. See also Glycosaminoglycan (GAG)
GAG polysaccharides
  in cartilage, 64
  in connective tissue proper, 63
Galectins, in cell adhesion, 18–19
Galvanic corrosion, of biomedical metals, 145
Gamma sterilization, orthopedic applications of, 113
Gelation, in sol–gel process, 171–172
Gel-based drug delivery systems, PC-containing polymers in, 116, 119, 120. See also Hydrogel entries
Gene delivery, PC-containing polymers in, 116, 119–122
Gene delivery systems, bioactive, 25–26
Gene expression
  drug conjugates and, 122
  growth factor signaling modulation by, 25–26
General corrosion, of biomedical metals, 144